FAIRY TALE SCIENCE

WRITTEN BY
SARAH ALBEE

ILLUSTRATED BY
BILL ROBINSON

Odd Dot · New York

FOR RUSTY DAVIS,
PHYSICS TEACHER AND FRIEND.
—S. A.

TO MY DAUGHTER, TILDA,
WHO INSPIRES ME EVERY DAY WITH
HER IMAGINATION AND CURIOSITY.
—B. R.

CONTENTS

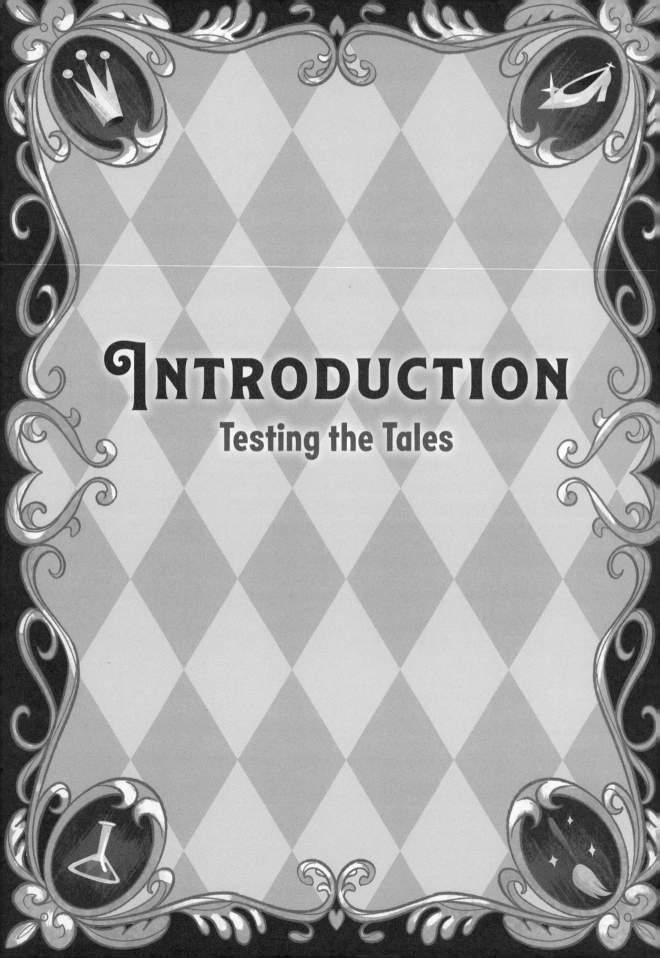

INTRODUCTION
Testing the Tales

Have you ever wondered
how a pair of glass slippers
could withstand an evening
of ballroom dancing?

Or if it's possible to spin
straw into gold?

How about if a wolf
could huff and puff and
blow a house down?

If these are the sorts of things you wonder about, why, then, you're in for a treat. Because in this book we'll investigate all these questions and more. You'll read fairy tales like a scientist. You'll ask questions like a scientist. You may even start *speaking* like a scientist, after you've pondered the scientific plausibility of particular plot points. See?

Here's how this book works:

Each chapter opens with a summary of a tale.

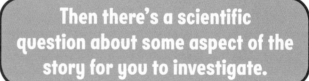

Then there's a scientific question about some aspect of the story for you to investigate.

And finally, each chapter has hands-on experiments and activities that relate to your investigation.

Our fairy tale inquiries will touch on many branches of science, including **physics**, **biology**, **chemistry**, **forensics**, **astronomy**, **geology**, **genetics**, **botany**, **anthropology**, **psychology**—and even the history of **alchemy**. (Those are all defined in the glossary, starting on page 215.)

There are a lot of big scientific words in this book. If you see a word or phrase in **bold purple type** and aren't sure what it means, look it up in the glossary at the back of the book.

Now here's your challenge:

After you've tried your hand at some of the experiments in this book, go back and read a few of the fairy tales again. What other, different scientific questions occur to *you*? Brainstorm ways to explore them. Plan and perform new experiments. Make improvements. Then ask more questions. You may be surprised at how much there is to investigate in these age-old stories about giants and witches and edible cottages and enchanted talking frogs.

Because sometimes science can seem truly magical.

What Were They Thinking?

But wait! you may be saying. *This is a nonfiction book! Fairy tales are made-up stories! How will fairy tale–based science even work?*

Many of the tales in this book were first told hundreds, even thousands, of years ago. The stories changed in the retellings from generation to generation. It took a long time before many were finally written down.

Some of the plots, improbable and fanciful as they may seem to us, reflect what people thought and feared, and the challenges they faced, at the time the stories were told. And there's often a thread that runs through some of the best tales, an echo of these thoughts and fears and challenges. It's fun to tug at these threads to examine what details in each tale might be scientifically possible, or which ones might have been based on actual historical events.

And remember: It wasn't so very long ago that most people believed in magic, and giants, and witches. When something out of the ordinary happened—a drought struck, or a cow died suddenly, or a family member fell ill unexpectedly—people from the past often attributed the event to a curse, or to sorcery, or to the alignment of the planets. And who could blame them? At the time some of these tales were told, an average-sized man stood a few inches over 5 feet tall. So is it any wonder that someone over 6 feet tall was considered a giant? Or that, in times when an alarming percentage of women died in childbirth, a woman who lived past forty, and who also knew how to mix healing potions from plants, was considered a witch?

And consider this: In pre-**penicillin** days, a person really could die after pricking their finger on a spindle. In the days before clocks and calendars existed, when time was a blur for most people, sleeping for a hundred years may not have seemed all that magical—it was just a way for the storyteller to indicate that a long time had passed.

Modern-Day Magic and Mystery

When you stop to think about it, have our beliefs really changed all that much? A lot of people today may not even realize how deeply science and magic coexist in our consciousness. Even the most scientific among us have been known to cross our fingers, make a wish before blowing out birthday candles, or pick up a lucky penny. Some of us even consider groundhogs to be reliable **meteorologists**.

So, whether you're a science fan, a fairy tale fan, or both, take out your notebook, pick up your pencil, grab your goggles, and let's get going.

A Note About Safety

Make sure to ask for an adult's help before performing any experiments that involve cutting, heating, lighting matches, using sharp objects, or handling hot substances. Be sure to put away materials, recycle what you can, and clean up any spills after you're finished.

Let's Meet The Key Players Behind The Tales!

Call them fairy tales, legends, folktales, or fables. However you describe them, they're all very old stories that were usually told out loud. Most have changed over and over again from one generation to the next.

WHO IMAGINED THEM?
More often than not, we don't know!

WHEN WERE THEY WRITTEN?
We don't know that, either!

WHY ARE WE LOOKING AT THE SCIENCE IN THESE FICTIONAL TALES?
Because it's fun!

We may not know who originally thought up most of the stories, but we know about some of the people who first wrote them down:

JACOB AND WILHELM GRIMM

These German brothers collected many old folktales, fairy tales, myths, and legends back in the nineteenth century and put them into a book. Their weird and violent versions of the stories have been terrifying small children ever since.

Charles Perrault was a French writer who collected and recorded a lot of now-famous tales to amuse his children, including "Little Red Riding Hood," "Cinderella," and "Sleeping Beauty." He published them in 1697. The book was called *Tales of Mother Goose*. Yep, some think he might actually be the original Mother Goose.

CHARLES PERRAULT

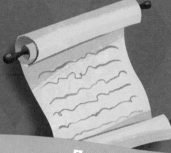

Aesop usually gets credit as the creator of many of the tales from ancient Greece that we now call *Aesop's Fables*. But we're not even sure he actually existed.

AESOP

Also known as *The Thousand and One Nights*, this group of Middle Eastern and Indian stories dates back at least a thousand years. The original tales probably had many authors and probably originated in many places.

ARABIAN NIGHTS

HANS CHRISTIAN ANDERSEN

Hans Christian Andersen lived in nineteenth-century Denmark and wrote a lot of deeply sad stories. The tales of his that we've included in this book— "The Princess and the Pea," "The Emperor's New Clothes," and "The Ugly Duckling"—are among his cheerier yarns.

Let's Meet the
STOCK CHARACTERS!

And now let's meet some standard fairy tale characters. We've updated some of them in this book, or at least called attention to how their roles reflect medieval-era thinking. Here's hoping these characters will continue to evolve as fairy tales are retold to modern audiences.

In your day and age, I'd be called a "scientist."

Most older women usually play the role of an angry fairy, an evil stepmother, or a carnivorous witch.

THE WICKED OLDER WOMAN

Fairy tale princes are usually not very picky about who they marry—their choice is often based on a young maiden's beauty, or her shoe size. If she happens to be comatose, well, so be it. In this book we'll cast a critical eye on some of these charm-boys.

THE HANDSOME PRINCE

Wolves are often the villains in fairy tales. This fear of wolves originated in medieval Europe, when much of the land was covered in dense forest. (Getting lost in the deep dark forest is another common theme.) Wolves symbolized people's fear of the wilderness. Real-life wolves don't deserve their villain status. Today these intelligent, social animals are endangered in many parts of the world.

THE BIG BAD WOLF

HELP!

Unfortunately, there aren't a lot of can-do female characters in fairy tales. Most young female characters tend to be helpless to get themselves out of trouble and usually need to be rescued.

THE DAMSEL IN DISTRESS

One Bad Apple

SNOW WHITE AND THE SEVEN DWARFS

A queen gives birth to a baby girl and names her Snow White. Then the queen dies. Her husband, the king, remarries. His new queen is Snow White's new, and wicked, stepmother.

Every day, the wicked stepmother asks her magic mirror to tell her who is the most beautiful woman in the land, and every day the mirror responds that she, the queen, is the fairest of them all. Meanwhile, Snow White grows older and becomes more beautiful day by day. We assume she has other qualities, such as a personality.

One day the wicked queen suffers a shock: The mirror tells her that she is no longer the fairest of them all. Snow White has knocked her down to second place on the beauty charts. Enraged, the queen sends for a huntsman and orders him to take the girl into the forest and

TALE ORIGIN

Versions of this tale appear in cultures around the world, including Ireland, Turkey, Iceland, Russia, Romania, Morocco, Libya, Mozambique, and Malaysia.
The first written version dates back to about 1750.

kill her. He partially obeys, but after he's brought Snow White deep into the forest he can't bring himself to do the deed, so he urges Snow White to run away. She passes a harrowing night in the forest, dodging sharp thorns and wild beasts. The next morning, she discovers a cottage. She lets herself in, tidies the place up, and then collapses with fatigue.

The cottage belongs to seven dwarfs. When they return home from work, they're surprised to find Snow White asleep in one of their beds, but they let her live with them and do all of their housework. She cooks and cleans for them while they're off at the mines every day, and all is well for a while. But then the evil queen's magic mirror informs her that Snow White is still very much alive. After several botched attempts to kill Snow White, the queen prepares a deadly poison, and disguises herself as an old peasant woman selling apples. She visits Snow White and convinces her to take a bite of a poisoned apple. Snow White falls down dead, or so it appears.

The heartbroken dwarfs can't bear to bury her, so they put her in a glass coffin. Along comes a handsome prince. When he sees the beautiful, albeit comatose, girl inside, it's love at first sight.

Disturbingly (given that he believes she is dead), he asks the dwarfs if he can take her, coffin and all, back home to his castle. They agree. In one version of the story, a servant of the prince stumbles while lifting the coffin, and the jolt dislodges the poisoned bite of apple from her throat and she revives. In another, the prince lifts the lid and kisses her on the lips, and she opens her eyes, which must have given everyone quite a start. Either way, she revives, and even though they've only just met, they get married and live happily ever after.

In some versions of the tale, upon learning of Snow White's marriage, the wicked stepmother has a temper tantrum and dies. In others, the stepmother is invited to the wedding and is forced to put on red-hot iron shoes and dance in them until she dies. At the wedding.

For the science question, let's turn to Snow White's near-death experience. The cause of her coma was, of course, the poisoned apple, which made her fall to the floor senseless, and which led everyone to think she was dead, because we assume that her heartbeat and breathing had slowed to an undetectable level. But then she fully recovered.

Let's consider whether it's possible for a person to make a full recovery after their pulse and breathing have slowed down or stopped. (And for other examples of slowed pulse and breathing, see page 146.)

EXPLORE the SCIENCE

Under what conditions might a person appear to be clinically dead, but then recover?

The Scientific Scoop: A Vital Sign

One of the first tests of a person's well-being is to check for the presence of a pulse. The pulse is a rhythmical throbbing in the arteries, which indicates that a person's heart is pumping blood around the body. Typically you feel for a pulse in a person's wrist or neck. In Snow White's case, you have to think that's one of the first things the dwarfs did upon finding Snow White lying unresponsive on the floor.

Find your own pulse by placing two fingers on the inside of your wrist beneath your thumb, or just beneath your jawline, close to your ear. Each beat that you feel is caused by your heart, which is a muscle that contracts and expands.

A slow pulse might suggest that a person is resting, or is possibly in a coma. A fast pulse suggests a person has just performed vigorous activity, or is somehow agitated. No pulse means something is seriously wrong— the person might have had a heart attack, or might be dead, or, in Snow White's case, might have taken a bite of an enchanted poisoned apple and is therefore only, er, *temporarily* dead.

But there are circumstances where a person's pulse may be so faint it can't be detected, or their heart might actually have stopped, and yet the person is still alive.

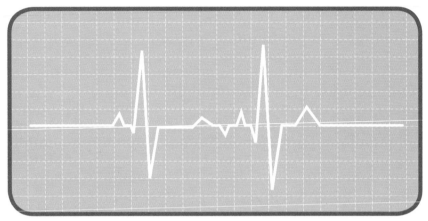

In 2015, a fourteen-year-old boy fell through the ice of a lake in Missouri and was under water for 15 minutes. When he was pulled to the surface he had no pulse and was not breathing. The medical team worked on him for an additional 30 minutes. The boy had been "**clinically** dead" for 45 minutes. But as the team kept up its frantic efforts to perform CPR (**cardiopulmonary resuscitation**), a pulse was detected. He eventually recovered.

How is this story—and others like it—possible?

Scientists have identified an evolutionary adaptation among humans and other **species** of animals that's known as the **mammalian diving reflex**. It enables animals to survive under water for extended periods of time. The reflex is triggered when a mammal's face comes in contact with cool water.

Receptors in the face send a message to the brain, which causes immediate physiological changes that help the body conserve oxygen. One of these changes is a drop of the heart rate. (For more about sensory receptors, see page 47.)

Before we test out the diving reflex, let's start by learning how to find our own pulse.

TRY THIS

FIND THE BEAT

Here's how to calculate the rate at which your heart is beating.

YOU'LL NEED:
- **A clock or watch**
- **A pen or pencil**
- **A notebook**

NOTE This experiment involves high-energy activity. Do this experiment only if you are in good health. If you feel faint, stop!

FOLLOW THESE STEPS:

1. Record your resting heart rate: Sit quietly for a few minutes to let your heartbeat settle into its resting rate.

2. Now find your pulse on your neck or on your wrist and count how many times you feel it beat within 10 seconds.

3. Multiply that number by six to calculate how many times your heart beats per minute.

4. Now do twenty jumping jacks, or run in place vigorously for a couple of minutes until you're good and winded.

5. Record your pulse again in a 10-second window.

6. Multiply that number by six and compare it to your resting heart rate.

7. Record your observations in your notebook.

What's Going On?

How did your heart rate change? Why do you think it did so? Just as exercise is good for strengthening your muscles, it's also good for strengthening your heart (which is also a muscle). Exercise elevates your heart rate and makes your heart muscle better able to pump blood throughout your body. The healthier your heart, the more blood it's able to push out with each beat.

CHALLENGE! What a Dive!

Want to test your own body's mammalian diving reflex? See how well it works.

YOU'LL NEED:

- **A heart-rate monitor or smartwatch**
- **A bowl of ice water large enough for your face (or a clean sink filled with cold water)**
- **A small towel**
- **A partner**
- **A pen or pencil**
- **A notebook**

 Please use extra safety measures when performing this experiment.

FOLLOW THESE STEPS:

1. Strap on the heart-rate monitor and sit quietly in front of the container of ice water for a minute or two while your heart rate settles.

2. Take a deep breath, then lower your face into the water.

3. Keep your face submerged for as long as you're able to hold your breath. (Try for 20–30 seconds.)

4. Have a partner observe (or even videotape) the heart-rate monitor.

5. Record your observations in your notebook.

What's Going On?

What happened to your pulse as your face was in the water? Why do you think your heart rate slowed, if that happened? How might what happened be beneficial to humans or large mammals?

PICK YOUR POISON

What kind of poison might have been in the apple?

It would have to be a **paralytic** poison—something that makes the victim unable to move. Snow White's breathing would have been so shallow, and her heart rate so faint, that she appeared to be dead. The poison in the apple was both fast acting and long-lasting. That will help us narrow things down.

Ready to play "Fun with Fairy Tale **Forensics**?" Here are some possible candidates for the poison used by the evil queen:

APPLE SEEDS

DEADLY NIGHTSHADE

CURARE

MANCHINEEL

POISON HEMLOCK

APPLE SEEDS

The seeds of ordinary apples contain trace amounts of a deadly poison called **cyanide**. But even if Snow White had taken a huge chomp out of the apple, and inadvertently swallowed a seed or two, the hard seed coating would protect her from whatever trace amounts of cyanide were inside the seeds she swallowed. And if she somehow managed to chew the seeds and compromise the seed coat, there would not be enough cyanide in a few apple seeds to produce any damaging effects. Snow White would need to eat about forty apples, core and all, to receive a fatal dose.

DEADLY NIGHTSHADE

A plant with a few deadly chemicals (called **alkaloids**), deadly nightshade has poisonous roots and dangerous berries that are bright, shiny, and dark. Symptoms of deadly nightshade poisoning include difficulty speaking, drowsiness, slurred speech, paralysis, coma, and, if you take too much—death. Deadly nightshade as the queen's choice for the apple is an interesting possibility. But one disconcerting effect of nightshade poisoning might disqualify it: Victims sometimes suffer from a thudding heartbeat, one that the dwarfs would certainly have been able to detect.

CURARE

Also known as *Strychnos toxifera*, curare is a plant-based poison made from boiling up the roots, bark, stems, and leaves of any of several tropical plants. It's a disturbing poison, because while it paralyzes victims—they can't move, their heartbeat slows, and their breathing gets shallow—they can still hear and feel everything. But one potential disqualifier for Snow White's apple is that curare is fairly harmless if you swallow it or breathe it in. To be effective, it needs to be injected.

POISON HEMLOCK

Don't be fooled by this plant that looks like parsley. *Conium maculatum* is a paralytic poison—the numbness attacks victims from the hands and feet first, and moves slowly inward and upward until it affects the lungs, and the victim suffocates. The most effective delivery system is to swallow the poison (rather than inject it or breathe it in). Could the evil queen have slathered poison hemlock juice on the apple? Possibly, although the smell might raise Snow White's suspicions. The smooshed stems of the poison hemlock plant have been described as smelling like a dead mouse.

MANCHINEEL

The evil queen could have given Snow White an apple from the *Hippomane mancinella* tree. Sometimes called by the cheerful name "beach apple," this pleasant-looking fruit grows in tropical places, where the trees can grow to be 15 meters (50 feet) high. All parts of the tree, including the flowers, the roots, and the fruit are highly toxic. Even standing under a manchineel tree when it's raining can raise blisters on your skin. So, the stepmother would have had to be wearing protective gloves when she handed over a poisoned apple from this tree, which probably would have tipped Snow White off to her nefarious plan.

Which poison do *you* think was the culprit?

The Kiss

Could something about the prince's kiss have helped reverse the effects of the poison in Snow White's system? A substance that counteracts the effects of poison is known as an **antidote**. Perhaps he was wearing a medicinal lip balm?

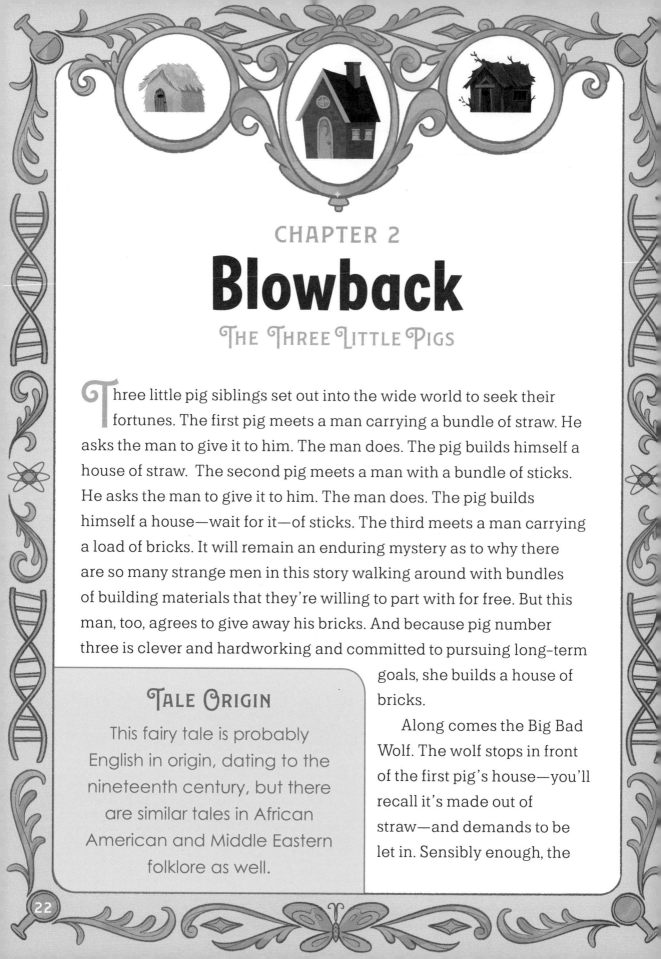

CHAPTER 2

Blowback

ᴛʜᴇ ᴛʜʀᴇᴇ ʟɪᴛᴛʟᴇ ᴘɪɢs

Three little pig siblings set out into the wide world to seek their fortunes. The first pig meets a man carrying a bundle of straw. He asks the man to give it to him. The man does. The pig builds himself a house of straw. The second pig meets a man with a bundle of sticks. He asks the man to give it to him. The man does. The pig builds himself a house—wait for it—of sticks. The third meets a man carrying a load of bricks. It will remain an enduring mystery as to why there are so many strange men in this story walking around with bundles of building materials that they're willing to part with for free. But this man, too, agrees to give away his bricks. And because pig number three is clever and hardworking and committed to pursuing long-term goals, she builds a house of bricks.

Along comes the Big Bad Wolf. The wolf stops in front of the first pig's house—you'll recall it's made out of straw—and demands to be let in. Sensibly enough, the

ᴛᴀʟᴇ ᴏʀɪɢɪɴ

This fairy tale is probably English in origin, dating to the nineteenth century, but there are similar tales in African American and Middle Eastern folklore as well.

pig refuses. The wolf huffs and puffs and blows the house in. The pig escapes next door to his brother's house—the one made of sticks. The wolf demands entry, gets denied, huffs and puffs, and blows down the stick house, too. The two pigs escape to the third pig's house, the one built of bricks. Once more, the wolf demands to be let in. But because she is clever and hardworking and committed to pursuing long-term goals—for example, staying alive—piggy number three refuses to let the wolf in. No matter how much huffing and puffing he does, the wolf is unable to blow down the brick house. In the end, the wolf tries to climb down the chimney, but gets tricked by the third little pig and falls into a big pot of boiling water. Ouchie.

How can strong huffing and puffing blow a house down?

The Scientific Scoop: Blow by Blow

First, a reality check: No wolf with an ounce of sense hunts its prey by blowing on it. And second, people build sturdy homes out of straw and sticks all over the world, and many of these homes can withstand all sorts of severe weather events, including large gusts of wind. It just takes some building know-how.

Still, wind can be a mighty powerful **force**.

To a scientist, a force is something that has the ability to change the motion of an object, often in the form of a push or a pull. **Gravity** is a force. **Magnetism** is a force. And yes, the mighty huff-puff of a Big Bad Wolf would count as a force, if there were such a thing as a talking wolf with a freakishly powerful lung capacity.

When wind pushes against a house, the house pushes back. These **opposing forces** illustrate Newton's Third Law of Motion, which states that for every action, there is an equal and opposite reaction. As long as the force of the wind and the opposing force of the house are equal, the house remains standing. But when the force of the wind becomes stronger than the opposing force, the house falls down.

Engineers have learned that during a hurricane or cyclone, even a house made of brick can fall down if the force on it is great enough. Wind rushing over the top of the house creates lower **pressure** above the house and can allow higher pressure inside the house to push the roof up and off. As a result, **structural failure** may occur—that is, the rest of the house can collapse.

Let's look more closely at opposing forces.

TRY THIS

USE THE FORCE!

Test the power of blowing air.

YOU'LL NEED:

- A blow-dryer
- One or more Ping-Pong balls
- Other lightweight materials, such as ribbon, an inflated balloon, or a toilet-paper roll
- A pen or pencil
- A notebook

FOLLOW THESE STEPS:

1. Switch the blow-dryer to the "cool" setting. Turn it on and point the air upward.

2. With your other hand, carefully position the Ping-Pong ball above the blow-dryer so it's in the stream of air.

3. Slightly move the blow-dryer from side to side. What happens?

4. Try adding a second or third Ping-Pong ball. Or try a balloon or a ribbon.

5. Think about the forces that are acting on the Ping-Pong ball. Record your observations in your notebook.

What's Going On?

The airstream from the blow-dryer pushes the ball upward, opposing the force of gravity, which pulls the ball downward. When the upward push and the downward pull are equal, the ball floats right where it is.

Now you may be wondering: *Why does the ball hover in place, rather than just blowing off and bouncing away?* Glad you asked. The air hits the ball at the bottom and whooshes around it in what physicists call a **slipstream**, before reuniting above the ball. The air in the slipstream moves faster, creating a lower pressure and a comfy little cocoon for the ball, which enables it to stay in place. This kind of airflow was first demonstrated by a Swiss scientist named Daniel Bernoulli, and is now known as the **Bernoulli principle**.

CHALLENGE! Huff-Puff Protection

Brainstorm and Design: Use different materials to build a house that can withstand some huffing and puffing.

YOU'LL NEED SOME COMBINATION OF THESE:

- **Drinking straws**
- **Craft sticks**
- **Sugar cubes**
- **Other building materials, including marshmallows, rubber bands, cardboard, toothpicks**
- **Scissors for cutting straws**
- **Serrated knife for cutting the craft sticks (ask a grown-up to help)**
- **Modeling clay**
- **Glue**
- **Tape (painter's tape is best)**
- **A pen or pencil**
- **A notebook**

AND ALSO: Fan or blow-dryer (Optional: Decorate the Big Bad Blow-dryer with felt ears and wiggly eyes)

FOLLOW THESE STEPS:

1. Construct a house with straws, sticks, or sugar cubes as bricks.

2. Use the modeling clay for sticking stuff together, and also, if you like, to make three little pigs.

3. Test the stability of your home by turning on the fan or pointing the blow-dryer at it. If a structural failure occurs, make changes to your design so it withstands the force of the blowing air.

4. Record your observations in your notebook.

Build, Test, and Redesign

Experiment by using other materials to reinforce your house. You may want to add marshmallows, rubber bands, cardboard, or toothpicks.

CHAPTER 3

Retweeting

The Nightingale

The emperor of China dwells in a magnificent palace, with gardens so vast even the gardeners don't know where they end. Beyond the gardens lies a huge forest. Beyond the forest, at the farthest end of the kingdom, lies the sea. The emperor learns that in one of the trees in this vast forest, close to the sea, lives a little gray bird called a nightingale. He is informed that the little nightingale sings so sweetly, the fishermen all stop what they're doing to listen to her song.

The emperor commands his minions to find the nightingale and to bring her back to the palace. This they do.

The bird's singing is indeed so beautiful that it makes the emperor weep. As thanks, he offers her lavish gifts—but gold and jewels are useless to her because, well, she's a bird. But because this is a fairy tale, the little bird does have the power of speech. She politely declines his gold and jewels, saying the mere knowledge that her music has brought tears to an emperor's eyes is all the gift she desires. The emperor orders a golden cage to be built for her, and commands twelve servants to wait upon her. Twice a day she is let out of her cage, tethered to a golden thread. She becomes the court songbird.

One day the emperor receives a gift from the emperor of Japan. It's an artificial nightingale, which looks a little like the live one, except that instead of having plain, gray feathers, the artificial bird is made of gold and encrusted with jewels. When it's wound up, it sings just like the real nightingale, although it can only play one song. But the emperor is charmed, and he commands his servants to wind and play the artificial bird over and over again.

Everyone at the royal court is so enchanted by the mechanical bird that no one notices the real bird fly out the window and back to her forest.

A year passes. And then one day, having been wound and played practically nonstop, the artificial bird pops a spring and stops singing. Royal mechanics manage to fix it, sort of, but in order to prevent its mechanical gears from wearing out, the artificial bird can only be wound up and played once a year at special occasions.

More time passes, and the emperor falls ill and is very close to death. He lies in bed, barely able to breathe. He opens his eyes and sees Death sitting on his chest, ready to take him. The emperor also sees the artificial bird next to his bed, and he commands it to sing to him one final time. But smart speakers with voice recognition technology are still many centuries in the future, and the windup bird remains silent.

Then suddenly, the emperor hears singing. The real nightingale has responded to his call and has flown from her tree in the forest to the branch just outside his open window. She sings her heart out for him. Her song is so sweet, the melody so lyrical, that Death Himself is reduced to pulp. He begs her to continue. She agrees, but on one condition: Death must let the emperor live. So Death compromises. She resumes her singing, and Death flies out the window in the form of a cold white mist. The emperor springs out of bed, all better. He promises that from now on the little bird will be free to come and go as she pleases, and she promises to sing for him often.

Tale Origin

This tale was published in 1843 by Hans Christian Andersen, who may have been inspired to write the story because he had a huge crush on a Swedish opera singer.

EXPLORE the SCIENCE

Could a mechanical device sound just like a living thing?

The Scientific Scoop: Natural or Not?

When the writer Hans Christian Andersen first published "The Nightingale" in 1843, audio recordings had not yet been invented. But windup musical devices *did* exist. From an engineering standpoint, these were ingenious gizmos. The windup device contained a small revolving cylinder attached to a steel comb that plucked tiny metal teeth. The teeth vibrated at certain **frequencies** and played a melody. But a windup nightingale would have sounded nothing like the real thing.

Nowadays, audio recording has become extremely sophisticated and can sound every bit as good as an in-person performance. Still, lots of people prefer to hear a live show rather than a recording.

What Is Sound?

Sound can be explained scientifically as waves that are created when an object vibrates and causes air **molecules** to bump into other air molecules. The **sound waves** create a disturbance in the air molecules that moves from one molecule to the next all the way to your ears.

This process is called **oscillation**. The larger the oscillations, the louder the sound. An object's **pitch**—technically, the **frequency** at which it tends to vibrate when hit, struck, plucked, strummed, or somehow disturbed—is also known as its **natural frequency**.

Sometimes the pitch-producing object will start to vibrate more vigorously if something nearby vibrates at a frequency that matches the object's natural frequency. This is called **resonance**. And sometimes, an outside force can resonate with an object's natural frequency so strongly, the vibrations (oscillations) get larger and larger and cause objects to fall apart.

Earthquakes can damage buildings severely when the frequency at which the ground is shaking "boosts" the resonant frequency of the buildings. Wind gusts have set up extreme oscillations in windows on huge skyscrapers, causing them to vibrate so strongly that the windows pop out. In 1940, wind whistling through the wires of a suspension bridge near Tacoma, Washington, caused the wires to vibrate at their natural frequency so violently that the bridge collapsed.

And get this: Soldiers marching over a bridge can match the bridge's natural frequency with their rhythmical footfalls and cause it to shake apart. That's why, back in the days when soldiers did more marching across rickety footbridges, they were told to "break stride," or not to march in step when they crossed the bridge.

Resonance can happen in music, too. Maybe you've heard that an opera singer can sing a note that's loud enough to shatter a wineglass? That's actually very hard to do, because even trained singers can't **amplify** the note (sing it loudly enough) to get the glass to vibrate as needed to break. But a *recording* of a person singing the pitch that matches the wineglass's natural frequency could make the glass shatter—if the **volume** is loud enough.

TRY THIS

YOU SHOULD TALK

Can a recording of your voice sound better than the real thing?

YOU'LL NEED:

- **A clean, 1-liter (1-quart) plastic takeout container, or something similar with high sides**
- **A smartphone or small audio-playing device**
- **A pen or pencil**
- **A notebook**

FOLLOW THESE STEPS:

1. Play a song or audiobook on the smartphone or audio-playing device.

2. While the song or audiobook is playing, put the device into the container. How does the sound change?

3. Experiment with the speaker end of the device facing up and facing down. Do you hear any differences?

4. Record your observations in your notebook.

Observe and Question

What happens to the sound as you turn the device upside down? Can you hear a difference in the quality of the sound when your device is inside the container? If the sound got louder, why do you think that happened?

KEEP GOING:

YOU'LL NEED:

- **A large, clean, head-sized bucket or bin**
- **A smartphone or other recording device**
- **A pen or pencil**
- **A notebook**

FOLLOW THESE STEPS:

1. Press "record" on your device and sing one of your favorite songs. Then stop the recording when you're finished.

2. Put the bucket over your head.

3. Hold the device close to your mouth, right beneath the bucket. Sing and record the same song you sang before, and try to sing at a similar volume and pitch as you did the first time. Stop the recording when you're finished, then take the bucket off your head.

4. Play back both versions of the song. How do the two recordings compare?

5. Record your observations in your notebook.

What's Going On?

Did your voice sound louder and "richer" from inside the bucket? When sound waves are contained in an enclosed area, such as the bucket, they are amplified, or made louder. Because they're in a contained space and can't spread out, the sound waves are more  concentrated. The reflection of the sound waves off the bucket material also enhances the quality of the sound. This principle may explain why you sound better when you sing in the shower.

CHALLENGE! > Pitch Perfect

Can you get a glass to oscillate at its natural frequency?

YOU'LL NEED:

- **A wineglass (Some glasses work better than others. A delicate, thin wineglass usually works best.)**
- **Dish soap**
- **Clean dish towel**
- **Vinegar or water**
- **A pen or pencil**
- **A notebook**

FOLLOW THESE STEPS:

 1. Wash the glass with hot, soapy water, and dry it with a clean towel. This experiment works best if the glass is "squeaky clean."

2. Wash your hands well. Natural oils on your skin can interfere with the oscillation.

3. Dip the tip of your finger in the vinegar or water. With gentle but steady pressure, trace a path around and around the top of the glass. Can you get the glass to "sing"? This may take some practice.

4. Record your observations in your notebook.

What's Going On?

Were you able to get the glass to produce a sound? That pitch you hear is the natural frequency of the glass. Does pouring in more water change the natural frequency of the glass?

Opera Glasses

Can you sing a note that's loud enough to shatter a glass? Try it! For this experiment to work, the note you sing must exactly match the pitch of the oscillating glass.

Put on some safety glasses and gently ping the side of the wineglass with two fingers. The sound you hear is the wineglass vibrating at its natural frequency.

Now prepare to sing that note *extremely* loudly—about as loud as a jackhammer.

Ready? Go ahead and give it a shot.

DID YOU SHATTER THE GLASS WITH YOUR VOICE?

If so, please contact the Metropolitan Opera immediately to schedule an audition. If not, don't feel bad. Even a trained singer, matching the correct pitch, would probably not be able to shatter the glass. But a *recording* of a trained singer, with the volume cranked way up, just might get the glass oscillating hard enough to shatter.

CHAPTER 4

An Open Secret

ALI BABA AND THE FORTY THIEVES

A poor man named Ali Baba is cutting wood deep in the forest one day when he hears voices. Believing he might be in danger of getting robbed, he shimmies up a tree to hide. Sure enough, it *is* a band of robbers—forty thieves, to be exact. He watches them come to a stop in front of a huge rock. The leader of the robbers utters, "Open Sesame!" The door swings open, revealing the entrance to a large cave. The thieves are inside for some time. When at last they emerge, the leader says "Close Sesame!" and the rock door closes. After they've gone, Ali Baba climbs down and repeats the same phrase. The door swings open, and he steps into a cave filled with magnificent treasure. He smuggles out some bags of gold, remembers to say "Close Sesame!" and heads home to his wife, telling her to keep the gold a secret.

TALE ORIGIN

This tale is part of a group of stories known as the *Arabian Nights*, collected from Indian, Persian, and Arabic folklore. The stories may have been published in tenth-century Persia, with "Ali Baba" and more added to the collection in the 1600s.

But the wife of Ali Baba's brother gets suspicious of the couple's sudden change in fortune. She tricks them into revealing the secret. Ali Baba's brother, Cassim, although already quite wealthy, decides that he, too, wants to steal some of the treasure. Reluctantly, Ali Baba gives his brother careful instructions. At the cave, Cassim remembers the "Open Sesame!" command and stuffs his bags with riches, but when he hears the robbers outside he panics and blanks on the exit phrase. The robbers find Cassim in the cave and kill him. They cut his body to pieces and leave it behind as a warning to anyone else who might be tempted to rob them.

A worried Ali Baba heads to the cave and finds his dead brother in bits and pieces. He smuggles the body (parts) back home for a proper burial, but he knows he's in trouble. Once the thieves return and find Cassim's body missing, they'll realize that their cave has been discovered and will come looking for whoever removed the dead man.

This is the moment when Morgiana, a clever slave girl in Cassim's house, enters the tale and saves the day.

Morgiana makes a big show of traveling around town and declaring that her master Cassim is gravely ill. Then she takes Ali Baba to visit a tailor. After they pay the tailor a sum of money, he agrees to their plan. In the middle of the night, they blindfold him and lead him to Ali Baba's house. There the tailor stitches together the bits and pieces of Cassim's body for burial, so it looks like he died a natural death.

Before long, the thieves come to town in search of whoever stole from them. They hear the tailor bragging about his tailoring skills. He boasts that he even stitched a dead body back together. One of the thieves blindfolds the tailor and gets him to retrace his steps, and the tailor finds his way to Ali Baba's house. The thief puts a big *X* on the door to remember which house it is. But Morgiana puts *X*s on the doors of a bunch of other houses, which thwarts the thieves. Another thief tries again with the tailor. This time, when the blindfolded tailor finds Ali Baba's house, the thief chips a piece of stone out of the front step to mark it. But Morgiana chips all the neighbors' steps, too. Foiled again! Finally the leader of the robbers himself accompanies the blindfolded tailor to Ali Baba's house and memorizes exactly where it is.

A few days later, the robber leader shows up at Ali Baba's house pretending to be an oil merchant in search of a place to stay for the night. He has with him many large ceramic oil jars, and Ali Baba, who seems to lack basic common sense, invites him in. Inside one of the jars there's oil, but in each of the rest of them hides one of the thieves. The plan? When the robber leader gives the signal, they're to bash their way out of their jars, murder Ali Baba, and steal back their treasure.

Morgiana comes to the rescue again. She coolly pours boiling oil into each jar, killing every one of the men hiding in the jars. Then, after dinner, as she's dancing in front of the robbers' leader, she pulls a hidden dagger from the folds of her garment and kills him right in front of a shocked Ali Baba. But after Ali Baba comes to understand all that she has done for him, he thanks her and marries her off to his son. We assume the couple lives wealthily ever after.

EXPLORE the SCIENCE

Is it possible for a voice to open a heavy door?

The Scientific Scoop: Voice Commands

In the story, the entrance to the thieves' cave was opened by magic.

In real life, computers can now convert speech to a digital command, which, yes, can open a heavy door, and it just *seems* like magic. How do computers "recognize" a voice command?

Computer programs are a set of instructions that a computer follows to complete a task. A human being writes the program, or step-by-step instructions, that tells the computer what to do.

When you speak, you create sound waves, which are vibrations that travel through the air. (See page 30.) A computer program can translate the sound waves into digital data by measuring the waves at precise and frequent intervals. If it recognizes what you're saying, the program can command the computer to, say, open a door. Today's speech recognition systems have grown ever more sophisticated and powerful. They can process complicated words and sentences, as well as the speech of people who have accents or who speak very quickly.

TRY THIS

BIG BANGS

If you can't actually see sound waves, how might you observe the effect they have on nearby objects?

YOU'LL NEED:

- **A large bowl**
- **Plastic wrap**
- **A handful of uncooked rice, lentils, sugar, or another small grain-like item**
- **A metal pot or cookie sheet**
- **A large utensil that can make noise, such as a metal soup ladle, a heavy wooden spoon, or two large pot lids**
- **A pen or pencil**
- **A notebook**

FOLLOW THESE STEPS:

 Tightly cover the top of the bowl with plastic wrap.

 Sprinkle a small handful of rice on top of the plastic wrap.

3. Hold the pot 1/2 meter (1 1/2 feet) or so away from the bowl and loudly bang on it with the spoon (or clang the pot lids together).

4. Why do you think the rice grains behave the way they do? Record your observations in your notebook.

What's Going On?

Did the rice jump around? The sound waves you created by banging on the pots made the plastic wrap vibrate, causing the rice to bounce up and down.

CHALLENGE! Use Your Senses

How do we use all of our senses to **navigate** through our world? The tailor was able to retrace his way to Ali Baba's house, despite having been blindfolded. To find his way, he would have had to remember sounds, smells, and the feel of the road.

> ## YOU'LL NEED:
> - **A short, 1–1 1/2-meter (3–5-foot) rope or sturdy string**
> - **A blindfold, such as a necktie or a torn strip of an old T-shirt**
> - **A partner**
> - **A pen or pencil**
> - **A notebook**

FOLLOW THESE STEPS:

1. With your partner, decide who starts out as the walker wearing the blindfold. The other starts as the guide.

2. The guide holds one end of the short rope and the walker holds the other.

3. The guide leads the blindfolded partner slowly and carefully on a walk. The guide should let the walker know when to start and stop by tugging once or twice on the rope. The guide should make sure to walk in a place that is safe and free of hazards like traffic or uneven ground, but which is full of sounds that could act as good audio cues for the blindfolded partner.

4. The walker should listen to sounds, notice smells, and use their free hand and their feet to feel their way as they walk.

5. Now switch roles and repeat steps 2, 3, and 4.

6. Try to recall how you used all of your senses to navigate while blindfolded. Record your observations in your notebook.

What's Going On?

Were you able to identify locations along the route? Which of your other senses did you rely on to find your way? When one of your senses is unable to perform its usual function, you use your other senses to compensate. Think about how a visually impaired person uses sensory clues to navigate their way through the world.

CHAPTER 5

A Royal Pain

The Princess and the Pea

This tale is short and, if you really think about it, mildly disturbing. A prince declares that he is ready to marry, and nothing but a true princess will do for him. But though he meets a lot of eligible, marriageable princesses, he feels there's always something a little . . . *meh* about each one. He moves back in with his parents. Then he mopes around the castle, worrying about how he can be sure the woman he decides to marry is a genuine princess and not just a sort-of princess.

One evening during a terrible thunderstorm, the royal family hears a knock at the door. The king opens it and beholds a drenched and bedraggled girl on the doorstep. She claims to be a princess. They offer her a room for the night. Before everyone heads off to bed, the queen sneaks into the guest bedroom, where their visitor will soon be sleeping. The queen lifts up the mattress. She places a pea

Tale Origin

Hans Christian Andersen wrote that he first heard this old tale when he was a boy. His version, retold here, was published in 1835, but similar tales were told centuries earlier in Sweden, Italy, Persia, and India.

underneath it. Then she replaces the mattress and also stacks twenty more mattresses on top of it, and piles twenty down comforters on top of all that.

The next morning, they ask their houseguest how she slept. "Terribly!" the girl replies. "I scarcely closed my eyes all night. There was something so hard in the bed that I'm black and blue all over my body."

Despite her bad manners—it's impolite to complain about your accommodations to people who have invited you into their home—the girl has convinced the prince that she's a real princess. Because, as everyone knows, only a true princess has skin so delicate that it would be bothered by a pea beneath all that padding. Now certain that she's a true princess, the prince marries her.

The two kind of deserve each other. He's a snob and a whiner, and she's a complainer and the world's most annoying houseguest.

The tale, short as it is, does raise an interesting scientific question. How does our sense of touch work?

Under what circumstances could you "trick" someone's senses?

The Scientific Scoop: You've Got Some Nerves

You feel with your skin, which contains specialized cells called **sensory neurons**. These neurons gather in signals from the outside world and transmit them to the brain. When you touch something with your finger, or feel cold or pain or a pea under the mattress, your sensory neurons convert what you feel into **electrochemical** impulses. Then they transmit these impulses from one neuron to another along a pathway that leads to your brain. Your brain then tells you how something feels—for instance, "hot!" or "lumpy mattress!" In response to the signals it receives, your brain activates **motor neurons**, which carry electrochemical impulses in the other direction and may cause you to pull a mound of down comforters off your bed, or move to a more comfortable part of the mattress.

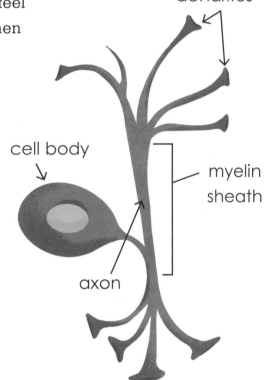

dendrites

cell body

myelin sheath

axon

But some parts of your body contain more sensory neurons than other parts, and sometimes your brain might get confused by the messages it receives.

TOUCHY-FEELY

Test out how well some of those sensory neurons work.

> **YOU'LL NEED:**
> - **Two dull pencils or a pair of chopsticks**
> - **A partner**
> - **A pen or pencil for writing**
> - **A notebook**

FOLLOW THESE STEPS:

1. Remove thick jackets or sweatshirts.

2. Turn your partner around so their back faces you.

3. Hold the pencils about 5 centimeters (2 inches) apart.

4. Gently touch your partner's back with either one or two pencils at the same time.

5. Ask your partner how many pencils they feel touching their back.

6. Try touching the pencils at different places on your partner's back, such as on opposite sides of their spinal column, near the base of their back, or on their shoulders. Can they feel a difference? Are different areas of the body more and less sensitive?

7. Turn your partner toward you. Have them close their eyes. Alternate touching one pencil or two to their palm. Then touch one or two pencils to their fingertips. Could your partner feel how many pencils were touching their palm? How about their fingertips?

8. Now switch places with your partner and repeat steps 4–7.

9. Record your observations in your notebook.

What's Going On?

Could your partner determine the number of pencil points accurately? How about you? Did the correct responses vary depending on where on their body they sensed the pencil points?

FLUFFY?

Your skin's touch receptors are not evenly distributed around your body. Some places, such as your hand, have lots of sensory neurons. Other places, such as your back, have fewer. If you're a ticklish person, you may know that certain parts of your body—such as the bottoms of your feet—are more ticklish than others.

CHALLENGE! | How Sweet?

Now try to "trick" your sense of taste.

> **YOU'LL NEED:**
> - **A mint candy or breath mint**
> - **Orange juice or an orange**
> - **A pen or pencil**
> - **A notebook**

FOLLOW THESE STEPS:

1. Dissolve the mint in your mouth until it's completely gone.

2. Take a sip of the orange juice or eat a section of the orange. How does it taste?

3. Record your observations in your notebook.

What's Going On?

Did the orange flavor taste like laundry detergent?

Your tongue is covered with little taste-sensitive cells. These taste receptors send messages to your brain that you're tasting something sweet, bitter, sour, or salty. There's a type of chemical in mint candy and minty toothpaste that somehow tampers with the taste receptors on your tongue and reduces your ability to taste sweetness. The chemical breaks up the fatty molecules on your tongue (called **phospholipids**) that ordinarily tamp down the receptors that taste bitterness, and actually *enhances* the bitter taste.

Mint tastes "cool" because menthol, the essential oil in mint, deadens the hot receptors in the mouth. Other essential oils, such as wintergreen, have similar **properties** and can help soothe aches and pains when rubbed onto the skin.

CHAPTER 6

Guess Again

RUMPELSTILTSKIN

We've all had moments when our parents have said embarrassing things about us in front of perfect strangers. But the opening scene of "Rumpelstiltskin" may rank as one of the biggest parent fails in all of literature. A father brags about his daughter to the king, boasting that she is so talented that she can actually spin straw into gold. So the king, who likes gold a lot, orders the girl to be brought to his palace to demonstrate her gold-spinning talents.

The king leads the girl (whose name we never learn) into a room with a pile of straw and a spinning wheel. He commands her to spin the straw into gold before the sun rises the next morning. If she fails, he will have her put to death.

After he leaves, the unfortunate girl stares at the pile of straw. Of course she has no clue how to change it to gold, and bursts into tears. Then in walks a little man. Upon learning why she is crying, he asks what she will give him if he spins the straw into gold. She offers him her necklace. He sits down at the spinning wheel and gets to work. Before dawn breaks, he finishes the task and leaves.

The next morning, the astonished king finds the room full of newly spun gold. He puts the girl into a bigger room, with even more straw, and tells her to spin *that* into gold by the next morning. Her penalty for failure: death. And the same thing happens: The little man appears and asks what she'll give him. This time she offers him her ring, and he agrees, and spins the straw into gold.

The next morning the king is delighted when he finds the gold. For a third time, he leads the girl to yet another, even bigger room, and commands her to perform the same task. But this time when the little man appears, she has nothing left to offer him. He makes her promise to give him her future firstborn child after she becomes queen. Probably deeming it a long shot that she'd ever become a queen, and also having no other option, she agrees, and the little man spins away.

After finding yet another room full of gold, the king is so pleased that he marries the girl. We're never told how she feels about this arrangement.

A year later, as the girl (who is now a queen) is rocking her new baby, the little man returns and demands that she turn over the kid as promised. She weeps and pleads and offers him all her jewels so she won't have to give her child to him. He appears to take pity on her, and says if she can guess his name by the end of three days, she can keep the kid.

On the first and second days when the little man appears, the queen runs through every possible name she can think of, to no avail. On the third and final day, the desperate queen sends a messenger out to the four corners of the kingdom to try to collect weird names she hasn't yet tried. The messenger reappears just before her deadline on the last day, and reports that he happened to encounter a strange little man dancing around a campfire. From his hiding place, the messenger observed that the man belted out a song about how the queen would never guess his name, tra-la-la, but that it's Rumpelstiltskin, tra-la-la.

TALE ORIGIN

Tales in which a character has to guess the name of a magical helper can be traced back 4,000 years, and exist in many cultures. This version, "Rumpelstiltskin," is best-known to English-speaking readers. But in other cultures he's called Whuppity Stoorie, Gwarwyn-a-throt, Purzinigele, Batzibitzili, and Panzimanzi.

So at the appointed time on the third day, the little man appears. She begins guessing. "Is it Tom?"

"No."

"Is it Dick?"

"No."

"Is it . . . Rumpelstiltskin?"

And Rumpelstiltskin is so angry she's guessed his name that he stomps his foot. He stomps it so hard he manages to cut himself in half, which is physiologically so impossible that we're not even going to consider it for scientific inquiry. Nope, we have another question to answer!

EXPLORE the SCIENCE

Can one thing be transformed into another without the use of magic?

The Scientific Scoop: Grow for the Gold

There's a long history of scientists who tried unsuccessfully to transform base metals, such as lead, into precious metals, such as gold. Performing this transformation was one goal of the ancient branch of natural philosophy now known as **alchemy**. The word comes from the Arabic *al-kimiya*. Alchemists never managed to create gold, but they did discover phosphorus, porcelain, toxicology, gunpowder, and the distillation process.

And modern scientists have discovered that it *is* possible to transform certain elements into other elements, though only in minuscule amounts. The process is called **transmutation**, and it helps if you have a particle accelerator, which costs about five billion dollars.

So, if you can't turn straw *into* gold, could you **extract** gold from straw? As it turns out, the roots of some plants have metal-absorbing capabilities. Scientists use such plants to remove toxic metallic elements from contaminated soil in a process called **phytoremediation**. Certain plants, known as metal **hyperaccumulators**, can vacuum up harmful elements like lead through their roots. Sunflowers, wheat, corn, and even common ragweed have proven to be good hyperaccumulators.

And yes, straw can indeed contain trace amounts of gold. Recently, scientists found a way to harvest gold that's been sucked up by plants. It's a process called **phytomining**. Scientists have sown certain fast-growing plants in areas near abandoned gold mines. The soil is treated with chemicals that help the plants' roots absorb tiny particles of gold left behind.

TRY THIS

PRESTO CHANGE-O

Yes—you, too, can change one thing to another! In this case, you'll change a flower's color (but no magic is necessary).

 NOTE This experiment may require an adult's help.

YOU'LL NEED:

- **At least four white carnations**
- **Four shades of food coloring**
- **Four glasses of water**
- **A small paring knife or scissors**
- **A pen or pencil**
- **A notebook**

FOLLOW THESE STEPS:

1. Add ten or so drops of food coloring into each of four different glasses of water. The darker the colors, the better.

2. With an adult's help, cut off the very end of each carnation stem at an angle.

3. Place one carnation into each glass of water.

4. Allow the flowers to sit for several hours or overnight. Observe your carnations. What happens to their petals?

5. Record your observations in your notebook.

What's Going On?

Why do you think the petals on your flowers changed colors?

Plants get most of the **nutrients** they need from air and water. (Plants don't actually need soil to live, but the nutrients in the soil help them grow and thrive better.) Water and nutrients are able to travel up the plant's stem through two processes: **capillary action** and **cohesion**. Capillary action is the upward pull on water as it gets sucked up by the plant's tube system (called its **xylem**). Cohesion is the attraction between water molecules, which causes them to stick to one another. That "stickiness" is known as hydrogen bonding.

Together, the forces of capillary action and cohesion help feed the plant, and in this case, pull the food coloring from the water to the petals.

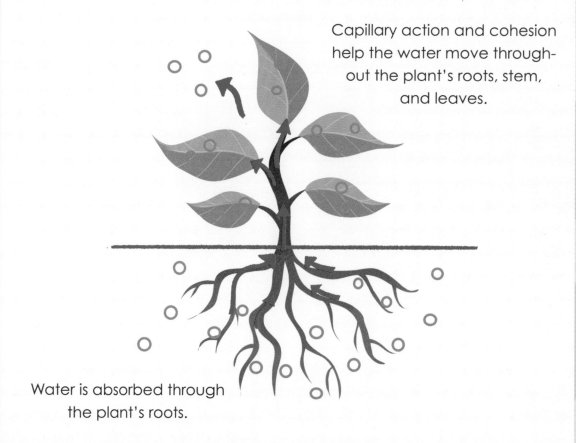

Capillary action and cohesion help the water move throughout the plant's roots, stem, and leaves.

Water is absorbed through the plant's roots.

CHALLENGE! On Your Marker! Get Wet! Go!

Rumpelstiltskin transformed straw into gold in a process known as, er, magic. In this challenge, you can transform one ink color into many different colors in a process known as, er, science.

YOU'LL NEED:

- **Several water-soluble or felt-tip markers, in colors like black, brown, gray, green, or orange (permanent markers won't work)**
- **Scissors**
- **Coffee filters**
- **Several small, shallow bowls or glasses**
- **Clothespins or binder clips**
- **Water**
- **A pen or pencil**
- **A notebook**

FOLLOW THESE STEPS:

1. Cut the coffee filters into strips—they should be about 2.5 centimeters wide by 10 centimeters long (1 inch wide by 4 inches long). Make one for each colored marker.

2. Fill each bowl or glass with water about 2.5 centimeters (1 inch) deep.

3. With a marker, color a thick line about 5 centimeters (2 inches) above the base of the filter strip.

4. With your pencil, write the color of the marker you used on each filter at the top of the strip. (After the filter gets wet, it can be hard to tell the original color.)

5. Set the strip into the water. Be careful not to let the colored part touch the water.

6. Secure the strip to the top of the bowl with a clothespin or binder clip.

7. Repeat steps 2 and 3 with the other colored markers. Place one filter into each glass.

8. Let the strips sit for 30 minutes or more.

9. Remove the strips and set them somewhere to dry.

10. How did the colors change over time? Make a list of the colors you found in each filter, and record your observations in your notebook.

What's Going On?

Chemists have a useful way of separating substances into their component parts. It's called **chromatography**, and it involves passing a **mixture** through a **medium** in which the components move differently. You just performed a version of chromatography in this experiment. The water-soluble colored ink is the mixture, and the coffee filter is the medium. Because some of the colors you chose are probably combinations of several colors, you were able to separate the colors out of the mixture. Different molecules moved up the filter farther than others.

The process of chromatography is useful for crime investigations. **Forensic scientists** separate the contents of complex mixtures into their chemical components and then analyze them. Chromatography can help investigators detect the presence of different kinds of drugs or poisons in a person's body.

CHAPTER 7

Prince Alarming

THE FROG PRINCE

In most versions of this tale, a princess is playing with her favorite toy, a golden ball. She throws it up and catches it over and over, but then she throws it too high and misses it, and the ball rolls into a well. She cries bitterly because she thinks it's lost forever, but then she hears a voice. It's a magical talking frog.

The frog offers to fetch her ball for her, and asks her what she'll give him in exchange. She'll give him anything, she replies. The frog lays out his conditions: He'll get the ball if she promises to let him come live with her at the palace, and eat off her golden plate, and sleep on her silken pillow. She considers the offer, and then agrees, because she really wants her ball back. Also, she's certain that there's no way this slimy **amphibian** could leave his watery home—even though he's magical and he talks. The frog disappears and reemerges with the ball in his mouth.

The treacherous girl grabs the ball and then runs home to the palace, leaving the frog calling feebly after her to wait for him. At supper that night, there's a knock on the door. The princess answers it, and when she sees the frog sitting on the stoop, she slams the door and goes back to the table. Her father asks what's going on, and when she tells him what happened, he insists that she honor her promise.

So the sulky princess has to let the frog hop up onto the table and eat from her golden plate. Then the frog reminds her that he also gets to sleep on her silken pillow. Because the king insists (thanks, Dad!), she carries the frog with two fingers to her bedchamber.

In one version of the story, the princess takes matters into her own hands. As soon as the frog hops onto her silken pillow, she picks

him up and flings him against the wall. Instead of going *splat*, the frog turns into a handsome prince who inexplicably declares his love for her, even though just a moment before she'd tried to smash him into subatomic particles. In another version of the story, the princess endures the frog's company for three nights in a row. On the morning of the third day, she awakens next to a handsome prince. He tells her an evil fairy had cast a spell on him that could only be broken by a princess letting him sleep on her pillow three nights in a row. She then falls in love with him and he with her. In all versions, the two live happily ever after.

Did you notice that the princess never has to *kiss* the frog? That's what most people remember from this story, but it doesn't actually happen, at least in the Grimm fairy tale version.

Let's back up to the very beginning of this story, and examine what might possess a princess—really anyone for that matter—to play with a ball made of solid gold right next to a deep well. It's a disaster waiting to happen. You don't need much scientific training to understand that a gold ball would sink to the bottom if it fell into the well. But it's more fun if you *do* know the science, so let's make **buoyancy** the focus of our scientific question and see why the ball would sink.

TALE ORIGIN

Stories that feature slimy suitors can be found all over Europe, including Poland, Ireland, Lithuania, ancient Rome, Turkey, Iceland, and Russia and may have been first written down in the thirteenth century. Tales of enchanted frogs appear in many cultures, including China, Sri Lanka, and Korea. This version is from the Grimm brothers.

What makes things float or sink?

The Scientific Scoop: Ball Game

Whether or not an object floats or sinks has to do with two things: its **buoyancy** and its **weight**. An object in water is pushed upward by the **buoyant force** of the water, and pulled downward by its weight. If the upward force of the water is greater than the weight of the water the object has pushed out of the way, the object floats.

A beach ball floats because it weighs very little, relative to its size. A solid gold ball sinks because its weight is greater than the buoyant force pushing up on it. When a beach ball floats, the buoyant force balances the force of gravity. When the gold ball sinks, gravity wins.

An enormous cruise ship floats because it pushes huge amounts of water out of the way, and the weight of that displaced water is equal to the weight of the ship.

Enter **density**. That's how much stuff is packed into a certain space. Salt water is denser than plain water. A gold ball is denser than a beach ball. And a peeled orange is denser than an orange with its relatively spongy, airy peel still on.

A ship stays afloat because the average density inside the ship is less than the average density of the water it's floating in. Engineers keep the amount of air space inside the ship less dense than the water, so that the ship remains afloat. If a ship's load gets too heavy, the ship can sink.

gravity

buoyant force

TRY THIS

A SINKING PEELING

Test the buoyancy of an orange.

YOU'LL NEED:

- **An orange with a thick peel**
- **A pot full of water**
- **A pen or pencil**
- **A notebook**

FOLLOW THESE STEPS:

1. Put the unpeeled orange into the water.

2. Remove the orange from the water and peel it.

3. Put the peeled orange into the water. What happens?

4. Record your observations in your notebook.

What's Going On?

An orange peel is full of hollow pockets filled with air, which makes the unpeeled orange less dense than the peeled orange. The air pockets in the peel allow the orange to float. It's like the orange is wearing a peel as a life jacket. Without its life jacket, the orange sinks.

Now, try the experiment with a thin–skinned fruit, such as a clementine. What happens?

GO FURTHER:

> **YOU'LL NEED:**
> - **A group of three or more people**
> - **A roll of aluminum foil**
> - **A pot or sink full of water**
> - **A cupful of pennies**
> - **A pen or pencil**
> - **A notebook**

FOLLOW THESE STEPS:

1. Give each person an equal-sized piece of aluminum foil. Aim for about 30-centimeter (1-foot) pieces.

2. Have each person form their piece of foil into a shape they think will float. Experiment with different designs and test them out in the water. You can modify your boat or try a different shape with another piece of foil.

3. Have each person set their foil boat on the water, then gently place pennies into their boat. Whose foil boat can hold the most pennies before sinking?

4. Record your observations in your notebook.

Build, Test, and Redesign

How did the shape of the boat affect the number of pennies it could hold? Experiment with different boat shapes. Does a boat with a flatter "hull" hold more pennies than a boat that sits deeper in the water? Does it make a difference to stack the pennies in one or two piles versus distributing them evenly within your boat? Why or why not?

 NOTE Recycle your aluminum when you're finished.

CHALLENGE! > Sink or Swim: To Float an Egg

Here's a way to demonstrate how density affects an object's buoyancy.

> ## YOU'LL NEED:
>
> - **A large jar or glass nearly full of water**
> - **A raw egg in its shell (at room temperature)**
> - **120 ml (1/2 cup) of table salt**
> - **A large soup spoon that can hold your egg**
> - **A pen or pencil**
> - **A notebook**

FOLLOW THESE STEPS:

1. With the spoon or two fingers, gently lower the egg into the water. What happens?

2. Remove the egg from the water.

3. Add the salt to the water and stir until it's dissolved.

4. Gently lower the egg into the salt water. What happens?

5. Pour out half of the salt water so the glass is only half full.

6. Tilt the glass and gently pour in some fresh water so the glass is nearly full again. Do not stir the water.

7. Gently put the egg back into the water. What happens?

8. Record your observations in your notebook.

What's Going On?

Why did your egg behave the way it did?

The egg probably sank when you put it into the plain water, and then floated when you put it into the salt water. Adding salt to your glass of water made the water denser. You increased the **mass** of the water without changing its volume (very much). In the last step, the egg was more dense than the plain water so it sank through it, but was less dense than the salt water so it sat on top of it.

CHAPTER 8

Home Sweet Home

HANSEL AND GRETEL

A brother and sister named Hansel and Gretel live with their father, who is a poor woodcutter, and stepmother, who is merely wicked, in a tiny cottage at the edge of a large forest. When famine strikes the land, the stepmother declares that they can't afford to feed four people any longer.

In a private conversation, the stepmother convinces Hansel and Gretel's father to take the kids deep into the forest and abandon them. Because they all live together in a tiny cottage, the kids overhear this private conversation. So when their father takes the kids deep into the forest, the kids are prepared. Hansel marks the trail with white pebbles, and they manage to find their way back home. The stepmother orders her husband to try again. The second time, Hansel drops a trail of crumbs, but the birds eat the bits of bread and the kids become lost in the woods.

TALE ORIGIN

This tale seems to have European roots dating back several hundred years, although versions of the story can also be found in Mexico, in both Spanish and Nahua cultures. This version is from the Brothers Grimm.

After wandering deeper into the forest, Hansel and Gretel come upon a cottage made of gingerbread and candy, with a roof made of cake and windows made of spun sugar. The hungry kids start nibbling on the house. What seems too good to be true, is. The children realize much too late that the house belongs to a witch.

The witch puts Hansel into a cage so she can fatten him up and eat him. Gretel remains free so she can do housework. Every day, the witch checks on Hansel. Because she has bad eyesight, she demands that Hansel stick out his finger so she can feel it to see if he's getting fatter. But clever Hansel sticks out a chicken bone instead of his finger. Every day, the witch feels the bone and declares he is still too thin.

Finally, she grows tired of waiting and orders Gretel to light the oven. Then she tells Gretel to poke her head inside to see if it's hot enough. Gretel is also clever, and she realizes the witch plans to eat her, too. So she pretends not to know how to work the oven. Exasperated, the witch demonstrates by poking her own head inside the oven. With a tremendous push, Gretel shoves the witch into the oven and closes the door, securing it with a bar. The witch is, quite literally, toast. After Gretel rescues Hansel, the kids fill their pockets with heaps of treasure and set out to find their way home. They discover the path and arrive at their cottage at last.

Their father suddenly develops a personality and tells Hansel and Gretel he is thrilled to see them back. He tells them that while they were away, their stepmother died. The kids dump the treasure out of their pockets and assure him he never has to chop wood or abandon them deep in the forest again. And they all live happily ever after.

EXPLORE the SCIENCE

Besides using pebbles and breadcrumbs, how else might someone find their way through a forest?

The Scientific Scoop: Getting Oriented

Because this tale was first told long before GPS was available, chances are Hansel and Gretel's best option to **navigate** through the forest would have been to use a **compass**. The magnetic compass was invented in China around 200 BCE, and was first used for navigation around the end of the twelfth century.

Here's how a magnetic compass works: Beneath the earth's outer core there's a moving liquid metal that creates powerful electric currents, and that causes the earth to act as an enormous magnet. A compass has a magnetized needle, which is mounted at the middle and can spin around freely. The magnetized compass needle has a north and a south end. Because opposite poles attract one another, when the compass is held level, the magnetized "south" end of the needle spins around until it points toward the earth's magnetic north pole. And once you know which way is north, you can figure out which way is south, east, and west.

True north (also known as the geographic North Pole) and **magnetic north** aren't quite the same thing, and compass needles point at slightly different angles depending on where you are on the earth. The earth's **magnetic poles** are slowly changing. Every half a million years or so, the earth's magnetic field reverses. Still, a compass would have been Hansel and Gretel's best bet.

TRY THIS

FIND A WAY

Make a compass for Hansel and Gretel to find their way through the forest.

 NOTE This experiment may require an adult's help.

YOU'LL NEED:

- **A small, uncoated metal paperclip**
- **A cork or a plastic bottle cap**
- **A serrated knife (for cutting the cork, to be done by an adult only)**
- **A strong magnet that you can hold in your hand**
- **A large, wide bowl of water**
- **A pen or pencil**
- **A notebook**
- **Optional: some colored nail polish or a permanent marker**

FOLLOW THESE STEPS:

1. If using the cork, have an adult cut a round piece about 1 centimeter (1/2 inch) thick, using the serrated knife.

2. Unbend the paperclip so that it is as straight as possible.

3. Optional: make a tiny dot on one end of the paperclip with the nail polish or marker to help you remember which end is magnetized.

4. Float the cork disk or the bottlecap in the center of the bowl of water.

5. Magnetize the paperclip: Holding it near the "colored" end, stroke the paperclip against the magnet 20–30 times along at least half of it, always in the same direction. This action helps the particles that make up the needle all point in the same direction.

6. Gently place your magnetized paperclip on top of the cork or bottlecap so that it is centered and both ends can spin freely.

7. The paperclip should swing around and point in a north-south direction. (You may need to give your cork or bottlecap a gentle bump with your finger if it drifts to the side of the bowl.)

8. Record your observations in your notebook.

What's Going On?

Can you tell which way is north based on your compass? The magnetized needle moves in sync with the earth's magnetic field and should spin around to point north. Which way is south, east, and west?

SIXTH SENSE

As you'll recall, Hansel and Gretel became lost after birds ate their breadcrumb trail. Have you ever wondered how *birds* navigate? How can birds leave your backyard in the autumn to fly south, and then find their way back to your yard the following spring?

Many birds have a special, sixth sense, as do dozens of other species of living things. It's known as **magnetoreception**, and it helps animals navigate, sometimes over thousands of kilometers. We're pretty positive that many species have it—from tiny **bacteria** to enormous whales. What we don't know is how it works, exactly.

There are two leading theories. The first: Perhaps the animals' bodies contain tiny magnetic particles that align with the earth's magnetic fields. The second: Perhaps the animals' bodies contain special sensory cells called **cryptochromes**, or light-sensitive molecules, that are able to help them orient themselves using both the sun *and* the earth's magnetic fields.

HONEY BEE (*Apis mellifera*)

It seems likely that honey bees may have the magnetoreceptor sensory cells in their abdomens that help them find food and determine the way back to the hive. Worker bees may use a combination of these sensory cells as well as visual landmarks and the angle of the sun.

HOMING PIGEON (*Columba livia domestica*)

These are a variety of the wild rock pigeons you see all over the world, but for over three thousand years, humans have bred the birds and trained them to deliver messages, thanks to the birds' amazing ability to find their way home—probably with magnetoreception. Homing pigeons were often used to deliver messages and mail in wartime.

Distances Traveled

Homing Pigeon:
1,770 kilometers
(1,100 miles)

Chinook Salmon:
3,057 kilometers
(1,900 miles)

Great White Shark:
5,500 kilometers
(4,500 miles)

Honey Bee:
12 kilometers
(7.5 miles)

CHINOOK SALMON *(Oncorhyncus tshawytscha)*

This species of salmon is born in freshwater rivers and streams but spends its adulthood in the ocean. After several years in the ocean, the salmon migrate hundreds of kilometers back to their original streambeds to deposit their eggs.

GREAT WHITE SHARK *(Carcharodon carcharias)*

In the eastern Pacific Ocean, great whites regularly migrate between Mexico and Hawaii. In other oceans, these majestic creatures migrate even longer distances.

MONARCH BUTTERFLY *(Danaus plexippus)*

Every fall, monarch butterflies travel from the northeastern US and Canada, across the North American continent, all the way to central Mexico. In the western part of the US, the butterflies migrate from Canada and northerly states to southern California. Many scientists believe the butterflies rely on the sun to navigate. But what do they do on cloudy days? It seems the butterflies also possess crypto-chromes. They may use this form of magnetoreception as a backup navigational tool.

LOGGERHEAD TURTLE *(Caretta caretta)*

The female loggerhead turtle may spend ten years at sea, but then she'll return to the very same beach where she hatched to lay her own eggs.

GRAY WHALE *(Eschrichtius robustus)*

Gray whales have the longest migration of any other mammals. Scientists think whales use different strategies to find food and family members, and to migrate vast distances, including some combination of magnetoreception and sonar (sound waves and echoes).

ARCTIC TERN *(Sterna paradisaea)*

These seabirds migrate annually from the North Pole to the South Pole—Greenland to Antarctica—and back again. Every year. That's a long way.

Loggerhead Turtle:
12,874 kilometers
(8,000 miles)

Gray Whale:
19,312 kilometers
(12,000 miles)

Monarch Butterfly:
8,047 kilometers
(5,000 miles)

Arctic Tern:
40,000 kilometers
(25,000 miles)

CHAPTER 9

Divide and Be Conquered

The Sorcerer's Apprentice

The main character in "The Sorcerer's Apprentice" is—wait for it—an apprentice. An apprentice is a person who learns an art or a trade from a more experienced professional in exchange for years of free labor. In this story, the apprentice is training to become a sorcerer, also known as a wizard.

The story was first published in 1797 as a fourteen-stanza poem, written in German by Johann Wolfgang von Goethe (pronounced, more or less, *GER*-tuh). It's been retold and expanded several times, but here's the general idea: The sorcerer departs the workshop and leaves the young apprentice behind to tidy up. The apprentice's first task is to wash the floor, but indoor plumbing hasn't been invented yet, and he doesn't feel like lugging buckets of water from the river. So he casts a spell on the broom and gets it to carry the water for him. But because he's not fully trained to perform magic, he messes up the spell. The broom carries bucket after bucket from the river and dumps more and more water onto the floor. The apprentice has no idea how to reverse the spell. He can't make the broom stop.

In desperation, he grabs an ax and chops the broom in half. But then both halves of the broom start fetching water, at twice the speed. In a later version of the tale, each of the pieces of the broom continues to divide in half until there are countless brooms carrying countless buckets of water. Just as the apprentice thinks that all is lost, the sorcerer returns and utters a spell that sets everything right again. The sorcerer reprimands the apprentice, reminding him that it's important to follow the rules and do as he's told and to leave powerful spells to the experts.

While a modern reader may root for the apprentice and applaud him for coming up with an innovative way to get his work done more efficiently, that takeaway was probably not shared by readers when this was first written. At that time, the ruling class was all-powerful, and lowly employees were expected to obey their superiors. But let's turn our attention to the broomstick, and look into whether something could be cut in half and become *two* somethings.

TALE ORIGIN

There are many variations of this tale type featuring a sorcerer and his pupil, and it may have originated in ancient Greece in about CE 170. Other versions have been found throughout Asia, Europe, Africa, and the Americas.

Can an organism be cut into multiple parts and function as independent organisms?

The Scientific Scoop: Heads or Tails?

Maybe you've heard that earthworms can be cut in half crosswise and that both halves can survive and become two separate worms. That's not actually true. Earthworms have a head end and a tail end, and while it may be possible for the head end to regrow a tail and survive, the tail end of the worm will die if it's separated from the head.

But other animals can perform amazing feats of **regeneration**. Zebra fish can regrow fins if they lose them. A lizard can "drop" its tail if it's caught by a predator, meaning it can break it off in order to escape. The lizard can regrow its tail later. Disturbingly, the segment of broken-off tail writhes around for a few minutes, which distracts the predator so the rest of the lizard can get away.

And there actually are creatures that can regrow major parts of their entire selves if they happen to get cut in half or cut to shreds: sea stars, sea cucumbers, and, a personal favorite of neuroscientists, the planarian flatworm—specifically the species known as *Schmidtea mediterranea*.

leopard gecko

This unlovely creature can regrow missing cells, organs, and even its *head* should it happen to lose it. The small freshwater worm can re-form from tiny slivers of itself. And—get this—if the worm is decapitated and grows a new head, it can *remember its surroundings*. Scientists aren't entirely sure how the worms manage to regrow new heads that can remember past life experiences, but it appears the worms have signaling pathways in places other than their heads. Signals can be sent along these pathways from one cell to another, and these signals are somehow able to transmit memories.

How can animals regrow their body parts and make new copies of themselves? They do so by using a remarkable yet minuscule part of the body called a **stem cell**. Stem cells are unspecialized—that means they have not yet been programmed to perform a single job in the body—and they have the ability to develop into any type of cell or tissue that a growing body requires. We all begin life with stem cells—that's why we can grow arms and legs and brains—but after our stem cells develop into muscle cells, skin

Totipotent stem cell

cells, brain cells, and other specialized cells, our bodies have fewer and fewer stem cells. But some animals have the unique ability to hang on to a special kind of stem cell—they're called **totipotent stem cells**—for their whole lives. Totipotent cells can grow into any kind of cell in the adult body.

Because these animals are able to hang onto these totipotent stem cells, they have the ability to regrow limbs, heads, and, well, all new selves.

Scientists are hopeful that one day soon, we'll be able to use what we've learned from these remarkable animals to enable humans to heal wounds and regrow body parts they've lost.

MY, HOW YOU'VE GROWN BACK!
SOME ANIMALS THAT CAN REGENERATE

ANOLE LIZARD
class: **Reptile**
It can regrow its tail.

VERTEBRATES

RED-SPOTTED NEWT
class: **Amphibian**
Among other parts, it can regrow its heart.

AXOLOTL
class: **Amphibian**
It can regrow complex body parts, like arms, legs, tail, heart, lungs, ovaries.

PLANARIAN FLATWORM
class: **Rhabditophora**
It can grow back its entire body from a speck of tissue.

SPIDER
class: **Arachnid**
It can regrow a leg.

INVERTEBRATES

MOON JELLYFISH
class: **Scyphozoa**
It can seal up a hole in its body and grow new limbs.

SEA SPONGE
class: **Demospongiae**
It can reconstruct its entire body, even if it's broken into tiny pieces.

SEA CUCUMBER
class: **Holothuroidea**
It can expel many of its internal organs out of its butt to scare off predators. (The organs regenerate in a few weeks.)

CHALLENGE! | Doubling Troubles

In some retellings of the tale, the apprentice's broomstick divides in half and becomes two broomsticks, and then those broomsticks divide themselves again and become four broomsticks, and then eight, and on and on until there are countless broomsticks fetching and pouring water. The term for the rate at which those broomsticks double in number again and again is known as **exponential growth**. Here's a way to understand how the rapid doubling works.

> ## YOU'LL NEED:
> - **A checkerboard or chessboard**
> - **About 220 grams (1 cup) of dried rice, lentils, or other small grain**
> - **A pen or pencil**
> - **A notebook**
> - **Optional: A calculator**

FOLLOW THESE STEPS:

1. Put one grain of rice or lentil on the first checkerboard square.

2. On the second square, place double the number of grains from the square before (in this case, two grains).

3. On the third square, double the number from the previous square (in this case, four grains).

How many grains will you put on the fourth square? How about the fifth? How far can you get before there are too many grains to count or fit on a square, or before you get sick of losing count and having to start over again? Can you calculate how many grains you would have on the sixteenth square? (Hint: It's five digits!)*

Draw a diagram in your notebook that shows how the grains of rice would double in quantity each time.

What's Going On?

How long did it take you to lose count? By doubling the number of grains each time you move to a new square, you've demonstrated the concept of exponential growth. The greater the number you start with, the faster the numbers escalate into increasingly higher values.

*Answer: 32,768

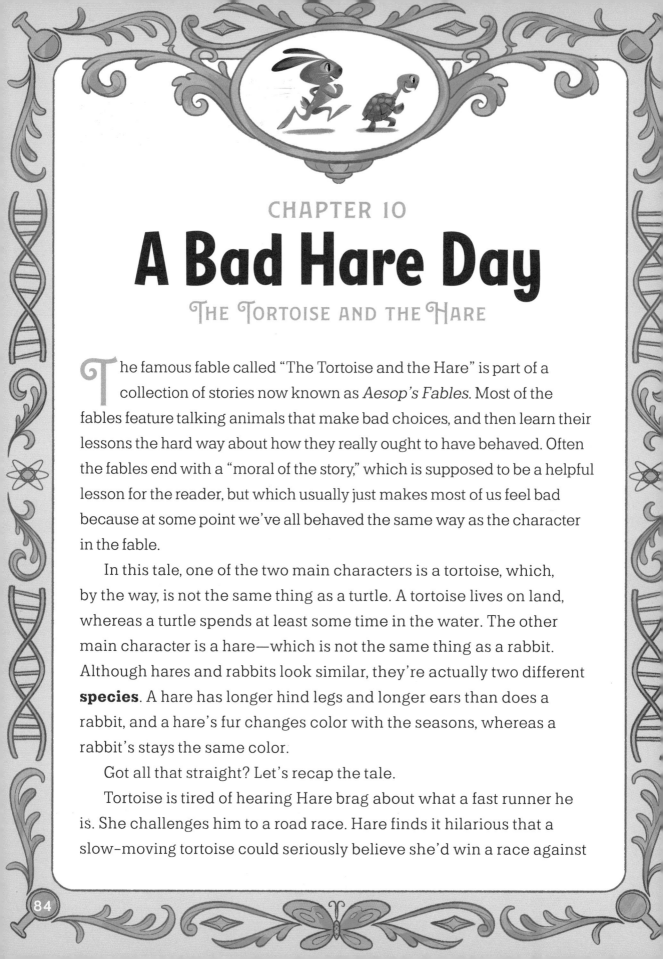

CHAPTER 10

A Bad Hare Day

The Tortoise and the Hare

The famous fable called "The Tortoise and the Hare" is part of a collection of stories now known as *Aesop's Fables*. Most of the fables feature talking animals that make bad choices, and then learn their lessons the hard way about how they really ought to have behaved. Often the fables end with a "moral of the story," which is supposed to be a helpful lesson for the reader, but which usually just makes most of us feel bad because at some point we've all behaved the same way as the character in the fable.

In this tale, one of the two main characters is a tortoise, which, by the way, is not the same thing as a turtle. A tortoise lives on land, whereas a turtle spends at least some time in the water. The other main character is a hare—which is not the same thing as a rabbit. Although hares and rabbits look similar, they're actually two different **species**. A hare has longer hind legs and longer ears than does a rabbit, and a hare's fur changes color with the seasons, whereas a rabbit's stays the same color.

Got all that straight? Let's recap the tale.

Tortoise is tired of hearing Hare brag about what a fast runner he is. She challenges him to a road race. Hare finds it hilarious that a slow-moving tortoise could seriously believe she'd win a race against a speedy hare, but he agrees to the contest.

Fox measures off the distance, sets them up at the starting line, and gives the ready-set-go. The tortoise and the hare set out. Hare bounds away and is soon far ahead of Tortoise. To demonstrate just how harebrained this race is, Hare yawns broadly and lies down by the side of the racecourse. He's confident that he's so speedy, he'll have plenty of time to take a nap and still beat Tortoise across the finish line. He falls asleep.

From far behind, moving slowly but steadily, Tortoise eventually plods past the sleeping Hare.

When Hare finally wakes up, he realizes Tortoise is almost at the finish line. He sprints as fast as he can, but Tortoise has just crossed the line and has won the race. The moral: slow and steady wins the race. Also, don't brag, and don't lie down and take a nap in the middle of a race.

Tale Origin

While this popular version of the fable is credited to Aesop, there are many similar tales that feature a race between unequal opponents. Such tales appear in African American and in Native American cultures (including Blackfeet, Pueblo, and Ojibwa), as well as in Brazil, the Philippines, India, and Fiji.

EXPLORE the SCIENCE

What are the benefits of "slow and steady" versus "speedy"?

The Scientific Scoop: Slowpokes and Speedsters

Tortoises, turtles, and other animals that lug around heavy shells tend to be slow moving. But because they do so much heavy lifting, they tend to have good **endurance**—that means they can continue to move at a steady pace. And shells offer protection from predators, which is a handy benefit for animals that can't run away from danger.

RACE YOU TO THE GARDEN!

Animals that can move quickly, like hares and rabbits, tend to be lighter and are able to dash in short bursts. In other words, they tend to be **speedy**.

In the human species, speed and endurance can vary from person to person. Are you someone who can beat most people in a sprint (that is, are you speedy)? Or are you the sort that can hike up a steep mountain without taking breaks to rest and catch your breath (that is, do you have good endurance)? Your tendency toward one type or the other might indicate whether you have more "fast twitch" or "slow twitch" types of muscle fibers. In scientific terms, human and animal muscle fibers contain a protein called **myoglobin**. Myoglobin provides oxygen to your muscles. Muscle fibers with a high concentration of myoglobin are the slow-twitch type. Those with a low concentration of myoglobin are the fast-twitch kind. Everyone's muscles are made of some combination of these two types of fibers.

The same is true of animals in the natural world. Mammals like seals and whales that can hold their breath for a long time tend to have greater amounts of myoglobin in their muscles than other animals. The extra stores of oxygen help them remain underwater longer. Speedy animals like cheetahs, lions, and hares have muscles with more fast twitch fibers than other animals. They can run fast with quick bursts of energy, followed by long periods of rest. The fastest land animal on the planet, the cheetah, can run 112 kph (70 mph), but it can only sustain that distance for about 1/2 kilometer (1/4 mile).

Slower-moving animals, such as sloths and tortoises, tend to spend most of their waking hours moving at a slower speed for longer periods of time. Slow-moving animals can have an advantage over faster ones, because they require less energy and don't have to eat as frequently. And researchers have discovered that the slower, steadier animals travel a longer total distance over the course of their lifetimes than do speedier animals that sprint for short bursts and then take long rests. So, slow and steady *can* win the (lifelong) race.

Compared to most land animals, humans are excellent endurance runners, thanks to our upright posture and our ability to cool ourselves off by sweating. We're right up there with ostriches and camels.

Survival of the Slowest

The Great American Horse Race of 1976 involved a course of 5,600 kilometers (3,500 miles) through thirteen states, across the continental United States and over lots of rough terrain. Riders traveled across prairies, through deserts, and over mountains. Many valuable, speedy horses from countries all over the world entered, and many riders and their horses dropped out along the way. The winner—by a landslide—was a guy on a slow and steady mule. (A mule is the offspring of a male donkey and a female horse.) The mule's name was Lord Fauntleroy.

TRY THIS

THE AMAZING ALLOWANCE TRICK

How fast is your reaction time?
Let's see what you're made of.

YOU'LL NEED:

- **A crisp dollar bill**
- **A ruler**
- **A partner**
- **A pen or pencil**
- **A notebook**

FOLLOW THESE STEPS:

1. Hold a dollar bill vertically.

2. Have your partner open her thumb and index finger so that the middle of the dollar hangs in the space between them.

3. Tell your partner that she can keep the dollar bill if she catches it before it falls.

4. Drop the dollar. Did your partner catch it?

5. Now try it yourself with the ruler. Have your partner hold the ruler from the top edge, so the bottom edge is just above your open thumb and finger.

6. Tell your partner to drop the ruler while you try to catch it.

7. Try it a few more times. Position your fingers at the same place every time the ruler drops. Each time you catch the ruler, note the closest inch marker to your fingers and write it in your notebook. The shorter the distance you catch the ruler, the faster your reaction time.

8. Record your observations in your notebook.

What's Going On?

Because you are holding them vertically, the dollar bill and ruler experience very little **air resistance** and fall quickly. Generally, people with faster reaction times tend to have more fast-twitch muscle fibers.

CHALLENGE! Run with It

How can you determine whether a person has more fast-twitch or slow-twitch muscles? Test out this activity in groups of three or more.

NOTE Wear running shoes for this activity.

YOU'LL NEED:

- **A large running space, such as an indoor gym, outdoor track, or a playground**
- **A clock or watch**
- **A pen or pencil**
- **A notebook**

FOLLOW THESE STEPS:

1. Pick one person to be the timer for the whole group. The rest of the group will run a total of three laps, one lap at a time.

2. As each runner completes the first lap, the timer calls out the elapsed time. Each person records their lap time.

3. After a 15-second break, the group runs a second lap. Each runner records their second time.

4. After a 15-second break, the group runs a final lap. Each runner records their time.

What's Going On?

Did the same person win each lap? Did one person start out fastest but "lose steam" by the end? Did some people start out slower, but prove to have more endurance by the end of the run? Note that there are lots of other factors that determine how fast or how long a person is able to run besides the makeup of their muscle fibers. But it's fun to find out what your muscles might be made of.

CHAPTER 11

Stirring Things Up

STONE SOUP

A couple of weary and hungry travelers arrive at a village. Their clothes are tattered and their boots are worn out. The villagers see them coming and fear that the men are beggars who will want something to eat, so they quickly close up their cupboards and stash away their bushels of vegetables.

Sure enough, the two men knock on the door of a house and ask for a morsel of food. The villager tells them there's nothing in the house to eat. They ask at the next house, and the next, and receive the same response.

At the next house, they ask only if they can have a stone. Puzzled, the villager tells them they are free to pick up whatever stone they wish from his front yard. Then the men declare that they're going to make stone soup and ask to borrow a large pot. With growing curiosity, the villager loans them one. In the middle of the town square the two men lay a fire, fill the pot with water, and put it on to boil. Aware that the whole village is now watching them with interest, one of the men dramatically drops the stone into the simmering water. One by one, baffled villagers come out to see what the men are up to.

"We're making stone soup!" announces one of the men.

"It will be delicious," adds the other. "And yet—how much better our soup would taste if only we had a bit of salt and pepper."

Someone brings them salt and pepper. Then one of the men reflects wistfully that their soup would taste so much nicer if they had a chopped onion or two. Someone brings them some onions, and into the pot the onions

go. Someone else brings carrots. Others bring potatoes, cabbage, and meat. More and more villagers suddenly "discover" these things in their cupboards and gardens and everything is chopped up and added to the pot.

The smell of simmering soup wafts through the streets. Practically everyone in the village has now joined the two travelers in the village square, and they've brought bowls, spoons, napkins, bread, and butter. At last the two travelers announce that the stone soup is ready. One of them fishes the stone out of the pot and sets it aside. Everyone in the village enjoys the feast with the two travelers. The stone soup is delicious.

There's no magic in this famous fairy tale. To a social scientist, the central theme explores how clever advertising can get people to open their wallets (or cupboards) for a product they didn't think they wanted, and in so doing, contribute something useful for the collective benefit of many. To a chemist, the making of stone soup is an excellent demonstration of both a **physical change** and a **chemical change**.

Tale Origin

This tale seems to have originated in Europe, although in a Russian version the story features "Axe Porridge," and in a Swedish version the characters make "Nail Soup." Don't try those at home.

EXPLORE the SCIENCE

What are physical and chemical changes, and what do they have to do with making soup?

The Scientific Scoop: Soup Science

First, some terms to get straight. A **mixture** is something you make by combining two or more different things.

A physical change happens when a substance is changed from one state—**solid**, **liquid**, or **gas**—to another, without altering the chemical composition of the substance. For example, if a liquid boils and changes to **steam**, that's a physical change (called **vaporization**).

A chemical change is a change that results in the creation of a new chemical substance by making or breaking the bonds between atoms. When a mixture of two or more things can't easily be "uncombined," that means that the particles that make up the multiple parts have been rearranged to form a new substance.

If heat is added to a substance and it produces a smell (such as the aroma of a cake baking), that's a chemical change. If adding heat causes the substance to change color (say, if you were to toast a piece of bread)— that would be another example of a chemical change.

Making stone soup could produce both a physical change and a chemical change. The soup would emit steam if it got hot enough—a physical change. As a liquid gets hotter, its **molecules** have more energy and move around a lot more. When the soup boils, the molecules have enough **kinetic energy** (the energy of motion) to leave the liquid in the form of steam.

The bubbling soup would create a delicious aroma—and that's a chemical change. Boiling the soup ingredients might cause the liquid to change color from clear water to cloudy broth—also a chemical change.

Little Miss Muffet

Remember the nursery rhyme known as "Little Miss Muffet"? She sits on a tuffet, eating her curds and whey, before getting terrorized by a spider. Ever wonder what curds and whey *are*? Well, they're the product of a chemical change!

Curds and whey are what's produced when you transform milk into cheese. To make cheese, you need to separate the liquid from the solid part of the milk by adding an **enzyme** (something that speeds up the process of separation). Cheese makers often use an enzyme called rennet. The rennet causes the milk to "curdle," or separate into clumps swimming in liquid. The clumps are the curds. The liquid is the whey. Depending on the type of cheese they're making, the cheese makers might drain off the whey and press the curds into a larger solid (the cheese). And there you have it: an example of a chemical change.

A MIX-UP

Create your own chemical change!

> ## YOU'LL NEED:
>
> - **5 grams (1 teaspoon) of baking soda**
> - **45 grams (3 tablespoons) vinegar**
> - **A deflated balloon**
> - **Duct or electrical tape**
> - **A large and empty soda bottle**
> - **A pen or pencil**
> - **A notebook**
> - **Optional: A funnel**

FOLLOW THESE STEPS:

1. Carefully pour the baking soda into the bottle. A funnel is helpful for pouring.

2. Pour the vinegar into the deflated balloon.

3. Attach the vinegar-filled balloon to the mouth of the bottle, but keep the balloon flopped to the side so the vinegar doesn't spill into the bottle.

4. Tightly secure the balloon to the bottle with the tape.

5. Raise the balloon so the vinegar trickles from the balloon into the bottle and mixes with the baking soda at the bottom of the bottle.

6. Give the bottle a little swirl to mix the vinegar and baking soda together, then watch what happens.

7. Record your observations in your notebook.

What's Going On?

Did the balloon inflate? If so, you've created a chemical change. The vinegar (a liquid) and baking soda (a solid) mixed together and formed carbon dioxide (a gas). As the carbon dioxide tried to escape from the bottle, it inflated the balloon, which is your evidence that a chemical change occurred.

HOW MATTER CAN CHANGE

PHYSICAL CHANGES

 **MELTING**

 **FREEZING**

 **VAPORIZATION**

 **CONDENSATION**

CHEMICAL CHANGES

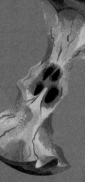 **ROTTING**

TARNISHING

 **BURNING**

RUSTING

CHALLENGE! What's Cooking?

Cooking can be a form of chemistry. When you make a cake, you're creating a mixture by combining **matter** (like butter, flour, sugar, and eggs). When you put it in the oven (add heat), the matter in the cake batter becomes a new substance: a cake!

You can also make your own stone soup. This is fun to make as a group project—each person can bring a different ingredient. Find a stone in a reasonably "natural" setting, as far away as possible from traffic, pesticides, or other chemicals. Ask a grown-up to help you scrub it with soap, soak it, and boil it so it's as clean as possible. (You can also run it through a dishwasher.) Add the stone to a pot of water, and bring to a simmer. Then add any combination of your favorite chopped vegetables, meat, herbs, seasonings, and other soup ingredients.

Don't forget to remove the stone before serving!

Have you created a physical change, a chemical change, or both? What evidence supports your findings? (If needed, go back to page 94 for help!)

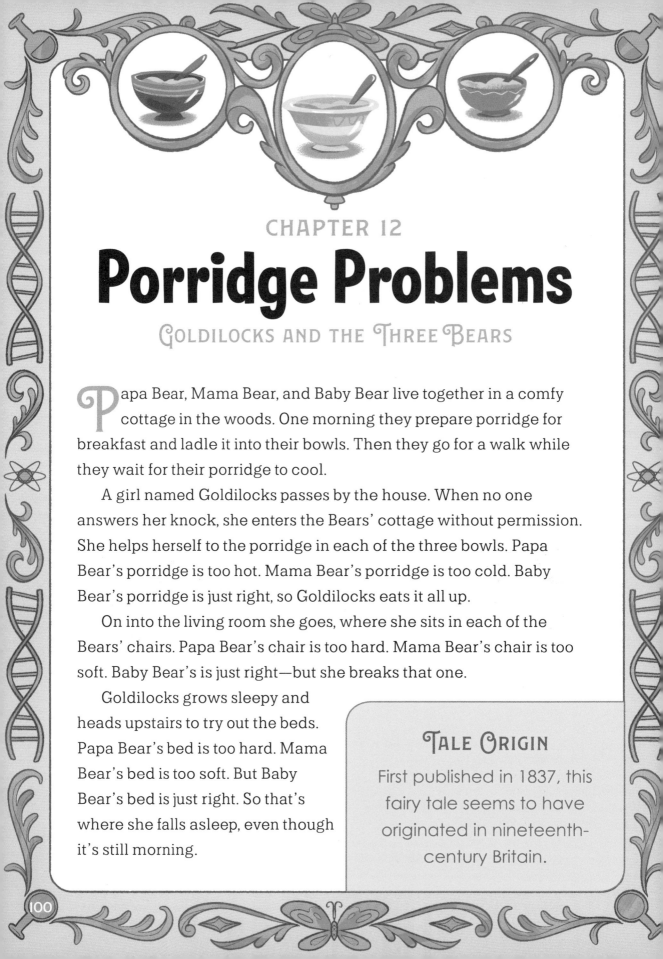

CHAPTER 12
Porridge Problems
GOLDILOCKS AND THE THREE BEARS

Papa Bear, Mama Bear, and Baby Bear live together in a comfy cottage in the woods. One morning they prepare porridge for breakfast and ladle it into their bowls. Then they go for a walk while they wait for their porridge to cool.

A girl named Goldilocks passes by the house. When no one answers her knock, she enters the Bears' cottage without permission. She helps herself to the porridge in each of the three bowls. Papa Bear's porridge is too hot. Mama Bear's porridge is too cold. Baby Bear's porridge is just right, so Goldilocks eats it all up.

On into the living room she goes, where she sits in each of the Bears' chairs. Papa Bear's chair is too hard. Mama Bear's chair is too soft. Baby Bear's is just right—but she breaks that one.

Goldilocks grows sleepy and heads upstairs to try out the beds. Papa Bear's bed is too hard. Mama Bear's bed is too soft. But Baby Bear's bed is just right. So that's where she falls asleep, even though it's still morning.

TALE ORIGIN
First published in 1837, this fairy tale seems to have originated in nineteenth-century Britain.

The Bears return. They find that someone has tasted or eaten up their porridge. They find that someone has messed with the chair cushions—and that Baby Bear's chair is broken. Mama and Papa Bear find that someone has rumpled up their beds, and then Baby Bear discovers Goldilocks asleep in Baby Bear's bed. Goldilocks awakens and is oddly surprised to find the homeowners standing there—in their own home. She dashes down the stairs and runs off.

It's a strange story. Why does Goldilocks feel she can enter the Bears' home as though she owns the place? The young juvenile delinquent is guilty of both breaking and entering *and* destruction of property. To say nothing of having terrible manners.

And here's another perplexing detail in the story: Remember how Baby Bear's porridge was warm ("just right") while Papa Bear's was too hot and Mama Bear's was too cold? How is that scientifically possible? Let's explore.

How can a larger bowl of porridge cool off faster
than a smaller bowl of porridge?

The Scientific Scoop: Some Like It Hot-ish

The idea that a medium bowl of porridge would cool off faster than a small bowl of porridge, of the same material and shape, violates Newton's law of cooling—which, in a nutshell, says that bigger things cool off more slowly than smaller things. So, according to Newton's law, Mama Bear's porridge should have been warmer than Baby Bear's.

Let's assume that all of the Bears' bowls are made of the same material (like wood, ceramic, or metal) and that they're all the same shape. Let's also assume that the big bowl contains more porridge than the medium bowl, and that the medium bowl has more porridge than the small bowl. A bowl with more porridge in it has more **thermal energy** (also known as heat).

But what really determines the rate that each bowl releases thermal energy—how quickly the porridge cools off—is the **surface area** of the porridge in the bowl. A wide, shallow bowl of porridge, having more surface area, will lose heat more rapidly than a deeper bowl with a smaller surface area. And yet—as the bowl size gets larger, the volume increases faster than the surface area. (For more on size and volume, see page 108.) So, though Mama Bear's bowl has a larger surface area than Baby Bear's bowl, it has more porridge in it, and therefore it has more thermal energy. So her porridge should not be cooler than Baby Bear's porridge. Let's explore.

A PORRIDGE PROBLEM

Let's determine whether those bowls of porridge violated the laws of physics.

> ## YOU'LL NEED:
>
> - **A 2-liter (8-cup) glass measuring cup**
> - **Small, medium, and large bowls of similar shape and material**
> - **Hot water**
> - **A clock or watch**
> - **Three cooking thermometers (or, if you have just one, repeat the experiment with each size bowl)**
> - **A pen or pencil**
> - **A notebook**

FOLLOW THESE STEPS:

1. Fill the small, medium, and large bowls with very hot water, stopping about an inch from the top for each bowl.

2. Insert a thermometer into each bowl and record the temperatures in your notebook.

3. Wait 5 minutes, then record the temperatures again.

4. Wait another 5 minutes. Record the temperatures a third time.

5. Record your observations in your notebook.

What's Going On?

How do your observations compare to what happens in the story? If your results don't match what happens to the three bears' porridge, how might you explain it?

CHALLENGE! More Fun with Bowls

How much time does it take for *all* the bowls of water to reach room temperature? Record the time in your notebook. What does that observation suggest to you about how soon Goldilocks entered the cottage after the Bear family had left it?

Experiment with using bowls made out of different materials. Does a metal bowl cool off faster or slower than a wooden bowl that's the same size and shape?

Try using two bowls that are shaped differently but that hold the same amount of water. One might be shallower and wider (with more surface area) and one might be taller and deeper (with less surface area). Record your observations in your notebook.

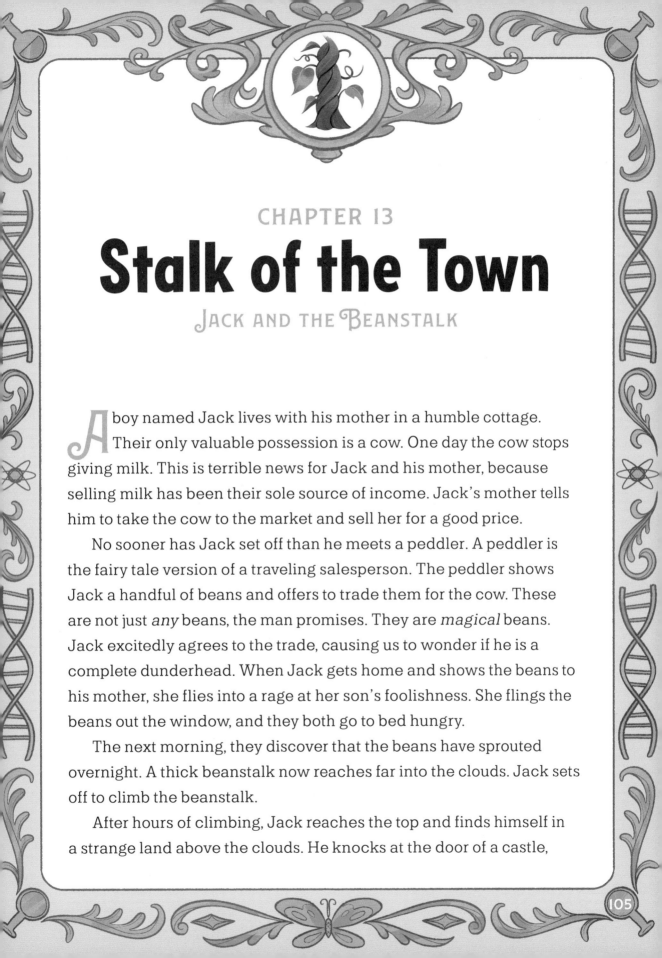

CHAPTER 13

Stalk of the Town

JACK AND THE BEANSTALK

A boy named Jack lives with his mother in a humble cottage. Their only valuable possession is a cow. One day the cow stops giving milk. This is terrible news for Jack and his mother, because selling milk has been their sole source of income. Jack's mother tells him to take the cow to the market and sell her for a good price.

No sooner has Jack set off than he meets a peddler. A peddler is the fairy tale version of a traveling salesperson. The peddler shows Jack a handful of beans and offers to trade them for the cow. These are not just *any* beans, the man promises. They are *magical* beans. Jack excitedly agrees to the trade, causing us to wonder if he is a complete dunderhead. When Jack gets home and shows the beans to his mother, she flies into a rage at her son's foolishness. She flings the beans out the window, and they both go to bed hungry.

The next morning, they discover that the beans have sprouted overnight. A thick beanstalk now reaches far into the clouds. Jack sets off to climb the beanstalk.

After hours of climbing, Jack reaches the top and finds himself in a strange land above the clouds. He knocks at the door of a castle, which he soon learns belongs to a giant. When the giant's wife opens the door, he begs her for some food, because he's hungry after all that climbing. After she feeds him, they hear the giant's footsteps. She tells Jack to hide in the woodbin, because her husband likes to eat boys for breakfast.

The giant stomps in. He sniffs the air and thunders, "Fee-fi-fo-fum! I smell the blood of an Englishman!" The wife tells the giant that his nose is mistaken and serves him his breakfast. Then she brings him his magic goose. From his hiding place, Jack watches the goose lay several eggs made of solid gold. When the giant falls asleep, Jack steals the goose and then escapes back down the beanstalk. He and his mother grow rich selling the gold eggs.

After some time passes, Jack decides to climb back up the beanstalk and reenter the giant's castle. As before, the giant fee-fi-fo-fums, eats his breakfast, and this time, takes out bags of money to count. When the giant falls asleep, Jack swipes one of the bags and skedaddles down the beanstalk.

During a third visit, Jack steals the giant's magic harp. But as Jack is creeping away, the harp suddenly develops lungs and yells for help. The giant awakens and chases after Jack, who hightails it down the beanstalk with the giant in hot pursuit. Jack yells to his mother to fetch the ax. He chops down the beanstalk with the giant still on it, and the whole thing—beanstalk and giant together—crashes down. The giant hits the ground and immediately dies. So Jack and his mother live happily ever after on the stuff Jack stole from the giant.

The disturbing moral? Crime pays. Or, it's okay to become rich by stealing from someone else, if that someone is not very nice.

TALE ORIGIN

We're not sure who came up with this version of the tale, but stories in which a trickster kid steals a giant's or ogre's treasure have existed for thousands of years, and in many cultures, including Japan, Spain, Romania, and Indigenous cultures from Nova Scotia to British Columbia.

EXPLORE the SCIENCE

If a giant could exist, what would that giant look like?
Would he be nimble enough to chase Jack down the beanstalk?

The Scientific Scoop: Shape-shifting

This tale contains a number of improbable plot points. First off, beans cannot **germinate** and grow as high as the clouds overnight. Second, a goose could not possibly lay several solid–gold eggs in a row. Three gold eggs would each weigh about 1 kilogram (about 2 pounds), and assuming the goose weighs about 3 kilograms (7 pounds), three eggs would be close to her entire body weight. And third, if an enormous beanstalk suddenly sprouted next to Jack and his mother's humble cottage and remained there for months, how could the neighbors not notice?

But let's focus our scientific inquiry on the giant, and ponder whether it's possible for someone of his size and shape to exist. Let's say Jack is 1 1/2 meters (5 feet) tall and the giant is 6 meters (20 feet) tall, or about four times Jack's height. A lot of problems occur if you try to scale up a person that much, while maintaining person-like proportions.

Here's what we can safely hypothesize: A giant that size would not have the proportions of an average human. Such an increase in size would put too much of a load on his bones. A giant who looks like a human except taller would break a thighbone on his first step. To support the enormous increase in weight (technically, volume) required for growing four times Jack's size, the giant would have to have much thicker legs relative to his body than Jack would. Having relatively shorter, thicker legs—in scientific terms, increased bone circumference and density—would make it very difficult for the giant to run after Jack and then climb down the beanstalk.

Because of his massive weight (again, technically, his massive volume), it's extremely unlikely that the giant would be nimble enough to climb down quickly, which is good news for Jack. By a similar principle, the base of a gigantic beanstalk would need to be extremely thick. It would take Jack hours to hack through it with an ax.

But the ending of the story *is* scientifically plausible. It's quite likely that the giant would die if he fell from, say, 9 meters (30 feet) up. And Jack and his mom would have quite a mess to clean up. *Blech.* (For more on falling objects, see page 129.) Let's look at a way to demonstrate how something's (or someone's) volume increases as it doubles in size.

TRY THIS

DOUBLE UP

Test to see what happens when you double the volume of a cube.

YOU'LL NEED:

- **Eight or more dice, or other cube-shaped blocks**
- **A pen or pencil**
- **A notebook**

FOLLOW THESE STEPS:

1. Select one die or block.

2. Using more dice or blocks, build a cube that is twice the volume (in other words, twice the length, width, and height).

3. How many total dice or blocks did you need?

4. Record your observations in your notebook

The mathematical formula for calculating the volume of a cube is

L × W × H (LENGTH × WIDTH × HEIGHT)

So in this case, it would be   2 × 2 × 2 .

Did your model look like this?

What if you were to double its volume again? How many dice or blocks would that require?

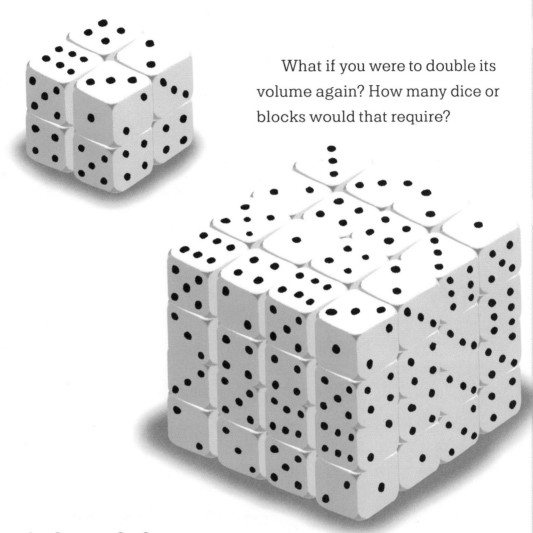

What's Going On?

As a cube gets twice as large in length, in width, and in height, its volume increases by a factor of eight. (Think how you went from one cube to eight cubes in order to double the volume of a single cube.)

Although it's easier to calculate the volume of a regular object, such as a cube, you can probably understand how rapidly the volume—and therefore the weight—of a large animal (or giant) would increase if it grew to quadruple its size. How might this principle explain why large, heavy animals have shorter, thicker legs?

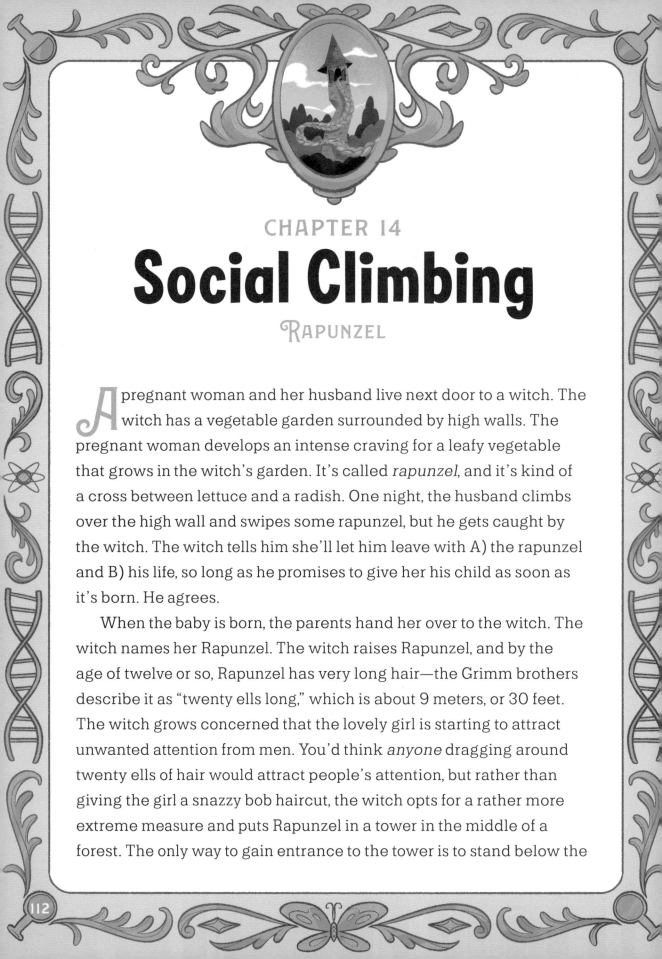

CHAPTER 14

Social Climbing

RAPUNZEL

A pregnant woman and her husband live next door to a witch. The witch has a vegetable garden surrounded by high walls. The pregnant woman develops an intense craving for a leafy vegetable that grows in the witch's garden. It's called *rapunzel*, and it's kind of a cross between lettuce and a radish. One night, the husband climbs over the high wall and swipes some rapunzel, but he gets caught by the witch. The witch tells him she'll let him leave with A) the rapunzel and B) his life, so long as he promises to give her his child as soon as it's born. He agrees.

When the baby is born, the parents hand her over to the witch. The witch names her Rapunzel. The witch raises Rapunzel, and by the age of twelve or so, Rapunzel has very long hair—the Grimm brothers describe it as "twenty ells long," which is about 9 meters, or 30 feet. The witch grows concerned that the lovely girl is starting to attract unwanted attention from men. You'd think *anyone* dragging around twenty ells of hair would attract people's attention, but rather than giving the girl a snazzy bob haircut, the witch opts for a rather more extreme measure and puts Rapunzel in a tower in the middle of a forest. The only way to gain entrance to the tower is to stand below the high window and bellow "Rapunzel, Rapunzel, let down your hair!" Rapunzel then twines the top part of her long braid around a hook and throws the other end of it down, and the witch climbs up to visit her.

One day a prince hears Rapunzel singing and falls in love with her . . . voice. He sees no way to gain entrance to the tower, so he hides and observes how the witch does it. The next night, he climbs up and meets Rapunzel. She's startled at first because he's the first man she's ever spoken to, but before long the two fall in love.

They come up with a scheme: Every day the prince will bring Rapunzel some silk so that she can fashion a ladder and eventually escape. Now, why she doesn't just A) ask him to bring her some scissors so she can cut off her own braid and tie it to the hook and climb down it to escape, or B) ask him to bring her a sturdy rope, is not clear.

It seems problem-solving is not Rapunzel's only challenge. Cluelessness is another, because one day when the witch comes to visit her, Rapunzel naively remarks that the witch is so much heavier to haul into the tower than is her boyfriend, the prince. Oops-a-daisy. In a rage, the witch cuts off Rapunzel's braid, and then banishes her from the tower. She deposits Rapunzel in a faraway desert.

Meanwhile, a few days later the prince drops by the tower and calls up to Rapunzel. The braid tumbles down, and up he climbs. But upon entering through the window, he encounters not his beloved, but an angry witch. In his shock and despair, he jumps out of the window and is blinded when thorns scratch his eyes.

The prince and Rapunzel wander around separately for a few grief-stricken years until at last one day, the blind prince hears Rapunzel singing, and they are reunited. Rapunzel's tears cure the prince's blindness, and the couple goes back to his kingdom and lives happily ever after.

TALE ORIGIN

The Grimm Brothers adapted this tale from a French story that was based on an earlier Italian version. We know it's a really old story. Variations of tales about "the maiden in the tower" have been collected in Europe, the Philippines, and ancient Persia.

How can human hair support the weight of a prince?

The Scientific Scoop: Insider Braiding

Hair is actually quite strong. We know this because we can measure what's called its **ultimate tensile strength**—that's how much load something can support before breaking. Hair is made up of tiny protein chains encased in an outer layer called the **cuticle**. Scientists have determined that a single strand of human hair is stronger than the same thickness of cast iron, aluminum, and copper. It would require only about 500 to 1,000 human hairs coiled together to lift an average 80-kilogram (175-pound) prince, and most teenagers have about 100,000 hairs on their heads. But before you rush off to try some heavy lifting with your hair, recall this detail: Rapunzel hoists the prince up by her *braid*. The multiple strands in braided hair would bear a prince's weight equally, which is a great example of a principle known as **parallel load bearing**.

Hair may be quite strong, but at the scalp, where it originates, the bond is weaker. You'll recall that Rapunzel had to wind the top of her braid around a hook before anyone tried to climb it, to avoid painful hair loss.

But a more concerning plot hole is how a girl of Rapunzel's age could possibly have hair 9 meters (30 feet) long. Hair grows an average of 15 centimeters (6 inches) a year, and the world record for the longest hair on a teenager is just under 2 meters (a little over 6 feet) long. Also, twenty ells worth of hair would weigh about 9.5 kilograms (21 pounds) so she'd have a serious neck ache lugging that around her tower room all day. One solution? Maybe she had hair extensions put in. Let's assume she did so, and that the extensions were natural human hair, so it doesn't mess up our science.

TRY THIS

HEAVY LIFTING

Test the ultimate tensile strength of a piece of hair.

 NOTE This experiment can be done in a group, where ideally one person has long hair.

YOU'LL NEED:

- A few strands of hair, at least 12 centimeters (5 inches) long (You may be able to collect strands from your comb or brush.)
- Tape
- A paperclip
- A plastic sandwich bag
- A pen, pencil, or chopstick
- A small pile of pennies or marbles
- Two raised surfaces (such as two stacks of books, shoeboxes, or the backs of two chairs side by side)
- A pen or pencil for writing
- A notebook
- Optional: A small kitchen scale

FOLLOW THESE STEPS:

1. Securely tape one end of a strand of hair to the middle of the pencil or chopstick.

2. Partially unfold the paperclip to form a "hook," and gently poke one end of it through the top of the plastic bag, close to the middle.

3. Securely tape the other end of the hair to the clip, so that the bag and clip are suspended from the pencil or chopstick by the strand of hair. It might take some practice to get both ends securely taped.

4. Lay the pencil or chopstick across the two raised surfaces so that the bag is not touching the surface of the table. You might want to tape down the pencil or chopstick so it doesn't roll around.

5. Begin placing your pennies or marbles gently into the bag. Keep adding pennies or marbles one at a time until the hair breaks.

6. Count up the pennies or marbles that one strand of hair was able to hold. If you have a scale, you can weigh them. Record your data in your notebook. (You might choose to create a table to organize your data.)

7. Experiment with different strands of hair. Are some hair types stronger than others? What happens if you double or triple the strands? Can you lift twice or three times as much weight? Compare your results to others in the group, and other types of hair. Is one color hair stronger than another? Did the length of the strand affect your results?

8. Record your observations in your notebook.

What's Going On?

A single strand of hair should be able to hold about 100 grams of weight (about 1/4 pound). If the prince weights 80 kilograms (176 pounds), how many pieces of hair would it take to lift him? If Rapunzel's hair has 100,000 strands, and they're braided together so they share the load-bearing equally, how heavy a weight could her braid support?

CHALLENGE! > Stresses on Tresses

After braiding her hair, Rapunzel was able to lift the prince, thanks to parallel load bearing. Here's a cool way to twist yarn into a handy rope and demonstrate that principle.

> **YOU'LL NEED:**
> - **Some balls of yarn in different colors**
> - **Scissors**
> - **Tape**
> - **A pen or pencil**
> - **A notebook**

FOLLOW THESE STEPS:

1. Cut six equal lengths of yarn—about 1 meter (3 feet) each.

2. Tie the strands together in a knot at one end.

3. Affix the knot somewhere—you can tie it to a doorknob, or tape it securely to a table.

4. Begin twisting the strands. Twist them in the same direction, starting at the knot end. Keep it tight and taut as you twist.

5. When you get to the end, hold it tightly. With your other hand, place a finger at the midpoint of your strand and loop the end back to the starting point, so your twisty-rope is doubled.

6. Hold the doubled strands tightly at the starting point, then open your fingers at the midpoint and let go. What happens?

7. Knot off the loose end.

8. Record your observations in your notebook.

What's Going On?

Six strands of yarn, doubled up to make twelve strands, have a much greater tensile strength than one strand. Try using this braided rope to lift different items. If you have access to hand weights, try tying it around different sized dumbbells. How much weight can it hold?

No wonder Rapunzel kept her hair in a braid!

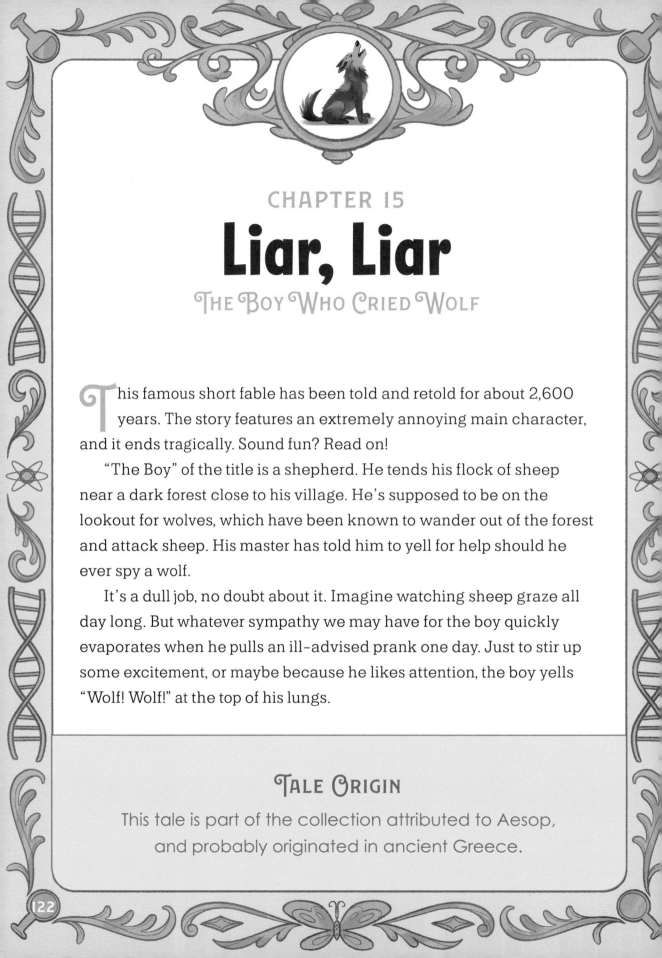

CHAPTER 15

Liar, Liar

The Boy Who Cried Wolf

This famous short fable has been told and retold for about 2,600 years. The story features an extremely annoying main character, and it ends tragically. Sound fun? Read on!

"The Boy" of the title is a shepherd. He tends his flock of sheep near a dark forest close to his village. He's supposed to be on the lookout for wolves, which have been known to wander out of the forest and attack sheep. His master has told him to yell for help should he ever spy a wolf.

It's a dull job, no doubt about it. Imagine watching sheep graze all day long. But whatever sympathy we may have for the boy quickly evaporates when he pulls an ill-advised prank one day. Just to stir up some excitement, or maybe because he likes attention, the boy yells "Wolf! Wolf!" at the top of his lungs.

Tale Origin

This tale is part of the collection attributed to Aesop, and probably originated in ancient Greece.

The villagers drop everything and run to help him drive away the wolf. They see no wolf. Instead, they find the boy rolling on the ground laughing at his hilarious practical joke. They don't share his amusement.

The next day he pulls the same stunt a second time. The villagers come running again, only to see the boy laughing hysterically . . . again.

And then one day an actual wolf shows up. It creeps toward the pasture, ready to fling itself upon the sheep. In genuine panic, the boy bellows "Wolf! Wolf!" but the villagers don't bother to run to help him because they think it's yet another of the boy's pranks.

The wolf carries off a bunch of the sheep, and we're left feeling sad for the sheep and mad at the boy. The moral of the story? Don't lie, because lying is wrong, and also because if you get a reputation as a liar, people won't believe you when you're actually telling the truth.

Not the most uplifting story, to be sure. Still, it's a useful tale for our purposes, because it raises some interesting issues having to do with fake news, social science, **criminology**, and **nonverbal behavior**.

How can you determine whether a person is lying or telling the truth?

The Scientific Scoop: Say What?

It's never been easy to tell for sure if a person is lying or telling the truth. In ancient China, suspected criminals were made to chew a mouthful of dry rice while they were being questioned. Then they had to spit out the rice. If some grains remained in their mouth, they were assumed to be guilty, because the mouth of a lying, nervous person produces less saliva, causing rice to stick to their dry mouth.

The lie detector, or **polygraph test**, was developed in the 1920s. Polygraphs measure and record things like blood pressure, pulse, and respiration while a person is asked a series of questions. But the test has never been 100 percent accurate—an honest person might be nervous when answering truthfully, and a dishonest person might be as cool as a cucumber when lying—so the results can't legally be considered in a court trial.

It seems the best way to tell if a person is lying is not so much by what they say, but rather by how they behave while saying it. In other words, we can study their body language, which social scientists call nonverbal behavior. A person's body language can provide clues to help determine whether or not the person is lying. Here are some behaviors to look for if you suspect someone might be lying.

The person:

Looks down or just avoids eye contact.

Shrugs or fidgets as they speak.

Crosses their arms and legs.

Covers their mouth as they speak.

Clears their throat or sniffs frequently.

Blinks their eyes or twitches their face.

Adjusts their clothing or glasses.

The ability to spot a possible liar is a useful tool for people in a number of fields, including detectives, lawyers, judges, trial juries, school principals, and people in earshot of bored shepherds.

TRUE OR FALSE?

Practice "reading" someone's body language.

> **YOU'LL NEED:**
> - **A partner, ideally someone you don't know very well**
> - **A pen or pencil**
> - **A notebook**

FOLLOW THESE STEPS:

1. Make up a list of ten questions with "yes" or "no" answers. They can be very straightforward, such as: "Are you ten years old?" or "Do you have any brothers and sisters?"

2. Ask your partner to look over the ten questions and decide in advance which ones they will lie about. Ideally, you should not know the answers to the questions you make up for your partner.

3. Sit opposite your partner and ask each question and record their answer in your notebook. Take your time moving from one question to the next. If you think the person might be lying, feel free to ask follow-up questions and note how they respond. Observe the person carefully, especially their nonverbal behavior. Write down the questions you believe they lie about.

4. After you've asked all ten questions, compare notes with your partner. How well did you determine whether or not they were lying or telling the truth?

5. Now switch roles and do the exercise again.

6. Record your observations in your notebook.

CHALLENGE! Two Truths and a Lie

Ever wonder why the guilt or innocence of someone accused of a crime is decided by a group of people? In the US, courtroom juries are usually made up of either six or twelve people. The right to a trial by a jury of one's peers is considered an essential right, and is written into our Constitution. The Founders believed that a jury of ordinary people making a decision about one person's fate was a far better option than, say, one cruel tyrant making all the decisions about *everyone's* fate. Also—added bonus—it turns out that a group of people working together to determine whether or not someone is lying can be quite effective (although by no means is it a foolproof system).

Give it a try yourself.

1. Form a group of three to five people.

2. Have each person write down three detailed stories or statements (three to four sentences each) about themselves. Two of the stories should be true. One should be made up. The more details each person can include, the more convincing their stories will seem. The other people in the group should not be familiar with anyone else's story. One possibility: "Great vacations I've taken." Another: "The best meal I ever ate."

3. Each person tells the group the three stories or statements about themselves. Pay attention to the person's gestures and body language as they tell each story.

4. Discuss your decision as a group and see if you are able to tell which story or statement is false.

5. Move to the next person in your group, who becomes the new storyteller.

6. Record your observations in your notebook and discuss your findings with the group.

CHAPTER 16

Falling to Pieces

HUMPTY DUMPTY

Okay, so if you want to get technical, this one isn't really a fairy tale. It's a nursery rhyme, and because it's only four lines long, you can read it here in full:

> **Humpty Dumpty sat on a wall.**
> **Humpty Dumpty had a great fall.**
> **All the king's horses, and all the king's men,**
> **Couldn't put Humpty together again.**

Why has this simple, four-line nursery rhyme been around for so long, and what makes it so catchy? For one thing, the name "Humpty Dumpty" is a really fun name to say, right up there with "Jack Spratt" and "Ethiopia." The rhyme has changed a little bit from the first time it was written down, back in 1797 in a book called *Juvenile Amusements*. Here's how the last two lines originally went:

> **Fourscore men and fourscore more**
> **Could not make Humpty Dumpty**
> **where he was before.**

TALE ORIGIN

This nursery rhyme originated in England.

You can find a few other early variations of the rhyme, but it's worth noting that none of them mentions that the character of Humpty Dumpty is an egg. In 1871, the writer Lewis Carroll (most famous for having written *Alice in Wonderland*) put the Humpty Dumpty character into his book called *Through the Looking Glass*, and it's there that Carroll described Humpty Dumpty as a large talking egg with clothes on.

The nursery rhyme does raise some interesting scientific and historical questions. For example, how might a horse, which lacks **prehensile** thumbs, even attempt to put a broken egg back together? Also, does Humpty Dumpty symbolize a real historical person who had "a great fall"? Ever since the time the rhyme first appeared, scholars with time on their hands have suggested different theories about who Humpty Dumpty is meant to represent. Does he symbolize certain kings that got overthrown or that died in battle?

It seems more likely that the original rhyme was meant as a riddle for kids—what can fall and break and not be put back together?—to which the answer was an egg.

All to say, we can't be sure if the original Humpty Dumpty was meant to be a large talking egg with clothes on, but let's roll with the egg interpretation, because egg experiments are really fun.

How can an egg have a "great fall" and remain in one piece?

The Scientific Scoop: Avoid Cracking Up

Whether or not a dropped egg will break or remain in one piece depends on several factors: the height from which it falls, the surface the egg lands on, and whether the egg has been protected with padding to cushion its **impact**.

If we're to go by the many illustrations of Humpty Dumpty's tragic fall, we can hypothesize that

A. he fell from a pretty high wall;

B. he landed on a hard surface—probably pavement; and

C. the snazzy clothes he was wearing did not provide sufficient padding to prevent him from breaking into pieces.

Let's consider the forces that were acting on him.

The most important force is gravity. As Humpty fell, gravity pulled him toward the pavement. The higher the wall, the faster he would have fallen and therefore, the greater the force of his impact with the ground.

Humpty was a large egg, possibly even human-sized, and therefore had a larger-than-your-average-egg mass. Thanks to a scientist named Galileo, who proved that all objects fall at the same speed, no matter what their size and weight, we can assume that Humpty hit the ground at the same **velocity** a regular-sized egg would have. *But* his large size applied more force to the ground, which in turn exerted a greater upward force on him. Because the **contact force** pushing back on him was greater than the strength of the molecules holding his shell together, poor Humpty experienced structural failure. In other words, he broke to pieces.

Galileo

Had he been wearing more padding, or landed on a softer or more sloped surface, or had his shell been less brittle, the outcome for Humpty Dumpty might have ended less tragically.

TRY THIS

SOFTEN UP

It's clear that a softer landing surface might have saved poor Humpty Dumpty. But would having a softer shell have helped save him? Here's a way to test that idea.

YOU'LL NEED:

- **A raw, uncracked egg**
- **A tall glass or jar**
- **About 250 milliliters (1 cup) of distilled white vinegar**
- **A sink**
- **A pen or pencil**
- **A notebook**

FOLLOW THESE STEPS:

1. Gently place the egg in the glass or jar.

2. Cover the egg completely with vinegar. Don't worry if it floats a little bit.

3. Leave the egg in the vinegar for 24 hours.

4. After 24 hours, carefully pour out the vinegar and rinse off the egg.

5. How far above a sink can you drop your egg and have it survive and bounce? Start at just a few centimeters (an inch or so) above the bottom of the sink. Work your way up to slightly higher heights. (Note: Always do this experiment over a sink.)

6. Record your observations in your notebook.

What's Going On?

Describe how your egg felt after you removed it from the vinegar. What happened when you dropped it from a height into the sink?

Eggshells are made of calcium carbonate. Vinegar is a kind of acid. By soaking the egg in the vinegar, you created a **chemical reaction**. (See pages 92 and 152.) The acidic vinegar began to dissolve the shell. The bubbles you see clinging to the eggshell are carbon dioxide gas, which is released during the chemical reaction.

KEEP GOING:

YOU'LL NEED:
- An egg
- A partner

NOTE This activity can also be done with water balloons.

Head outside for an egg toss with a partner. Stand about 2 meters (6 feet) apart, and gently toss the egg to them. Now take one step backward, and have your partner toss the egg back to you.

Did you catch it without the egg breaking? Think about *how* you caught it. You probably reached your hand out, caught the egg, and kept your arm swinging backward to minimize the **impact** (or force) of the egg hitting your hand.

Physicists have a term known as "impulse" to describe the direct connection between the time you take to slow down and stop the egg and the amount of force that's acting on the egg. If you'd held your hand up in a rigid position, the egg might have gone *splat* against your palm. You probably moved your hand along with the egg and slowed it down gradually.

Calculating the impulse of the egg involves multiplying the force on the egg by the time it takes to slow down and stop its **linear momentum** (that is, to catch it). A lot of force and a little time leads to a broken egg. When you catch the egg and take more time to slow it down and stop it, you've put less force on it and taken more time. Less force and more time leads to an unbroken egg.

CHALLENGE! > Under Pressure

Test the surprising strength of eggshells.

> **YOU'LL NEED:**
>
> * **Four eggshell halves, carefully washed and dried. You may need a couple more halves in case they don't have clean edges. Aim for equal-sized halves.**
> * **Masking tape or duct tape**
> * **Scissors**
> * **A collection of pantry foods, such as cereal boxes, cake mixes, sugar, and different types of cans.**
> * **A pen or pencil**
> * **A notebook**

FOLLOW THESE STEPS:

1. Wrap a piece of tape around the middles of each eggshell half.

2. Trim the taped edge with scissors so each eggshell has a straight-edged bottom.

3. Lay out the four half shells, dome up, to form a square. You might need to crack an extra egg or two in order to get four halves that are about the same height, and that are reasonably smooth and regular at the bottom edge.

4. Stack a cereal box or a can on top of the four half-shells. Gradually add more weight. How many objects can you stack before an eggshell cracks?

5. Record your observations in your notebook.

What's Going On?

What sorts of items did you put on top of your eggshells? Did the shells hold more than you expected?

If the experiment works well, you'll notice that eggshells can hold quite a bit of weight. Thanks to the domed shape, no single point in the curve supports the entire weight on top. The weight is carried down along the curved walls, and all four halves share the burden equally. (For more on eggshell strength, see page 213.)

Build, Test, and Redesign

Experiment with rearranging the eggshells. Try lining them up, pushing them closer together, or moving them farther apart. Can your shells support more weight? Record your observations in your notebook.

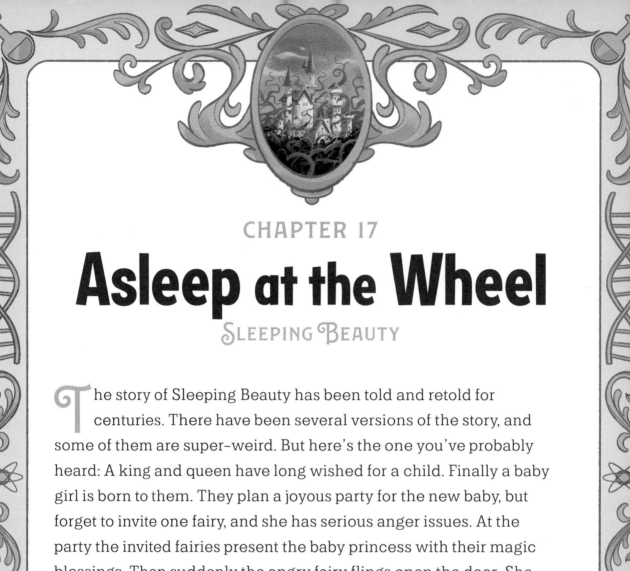

CHAPTER 17

Asleep at the Wheel

SLEEPING BEAUTY

The story of Sleeping Beauty has been told and retold for centuries. There have been several versions of the story, and some of them are super-weird. But here's the one you've probably heard: A king and queen have long wished for a child. Finally a baby girl is born to them. They plan a joyous party for the new baby, but forget to invite one fairy, and she has serious anger issues. At the party the invited fairies present the baby princess with their magic blessings. Then suddenly the angry fairy flings open the door. She hurls a curse at the innocent infant: On the princess's sixteenth birthday, the girl will prick her finger on a spinning wheel and die. The angry fairy flounces out.

Luckily there's still one fairy who hasn't yet bestowed her blessing on the princess. She steps forward and says that while she's not able to undo the angry fairy's curse, she can lessen its harshness. The princess won't die, but she *will* sleep for a hundred years.

The king has all the spinning wheels in the kingdom destroyed in an effort to avoid the curse, but of course it doesn't work. You'd think they'd warn the kid every day of her life to steer clear of spinning wheels, and that when her sixteenth birthday rolled around they'd lock her in her room to keep her safe, but fairy tales don't work that way.

On the princess's sixteenth birthday, the king and queen are "not at home." The princess goes wandering through the castle and comes upon a little tower room where she finds an old woman sitting at a spinning wheel. Sure enough, the princess pricks her finger on the

TALE ORIGIN

You can find similar stories dating back to ninth century China, but versions of this tale have been told all around the world. This tale is from the version written down by Perrault, which he based on a fourteenth-century Italian tale.

wheel and falls into a deep sleep. The rest of the castle inhabitants instantly fall asleep, too, and enchanted thorny brambles quickly grow around the castle, which is forgotten for a hundred years.

And then one day a prince passes by, and the thorny brambles magically part for him. He enters, and finds all of the sleeping people and animals. He finally arrives at the room in the tower and discovers the 116-year-old princess where she lies slumbering, and he is so charmed by her beauty, he kisses her. Her eyes open, and because this is a fairy tale, and fairy tale courtships tend to move quickly, the two instantly fall in love. They live happily ever after, although you have to think he will be spending a lot of time catching her up on the latest trends.

EXPLORE the SCIENCE

Could a person die from a prick to their finger?

The Scientific Scoop: Miracle Mold

In the days before penicillin and tetanus shots, which really wasn't very long ago, people understood the hazards of a small cut from shaving, a prick from a needle, or a scratch from a rose thorn. Today we'd consider these to be minor injuries. But at the time when "Sleeping Beauty" was first told, such trifling wounds could indeed lead to infection, blood poisoning, and death.

The nineteenth-century American writer Henry David Thoreau had a brother who nicked his finger with a razor. In a week, lockjaw, spasms, and convulsions set in—all indications of a **toxin** called **tetanus**—and then the brother was dead.

In 1915, the composer Alexander Scriabin died as a result of a sore on his upper lip that became infected.

Nowadays it would be extremely unusual for a person to die from a small cut or puncture wound, thanks to better hygiene, antibiotics, and the existence of **vaccines**, which help protect us from dangerous infections and illnesses. A vaccine is a tiny amount of a killed or weakened **virus** or **bacteria** that's injected into your body. Because the vaccine contains dead or weakened germs, it usually doesn't make you sick, but it does trigger your immune system to produce chemicals that can defend you, should you later be exposed to the real germs.

The tetanus vaccine was developed in the mid-1920s and became commercially available by about 1938. And penicillin, the first true **antibiotic**, was discovered in 1928, by Alexander Fleming. As far back as ancient Egypt, people had applied a concoction of moldy bread as an antiseptic for wounds, but Fleming was the first to isolate penicillin from mold, and to realize the drug's huge potential. It took several decades before scientists figured out how to produce penicillin in mass quantities.

The first-ever patient to be treated with Fleming's form of penicillin was Albert Alexander, who in 1941 scratched himself with a rose thorn. He became seriously ill with blood poisoning.

Alexander was treated with penicillin and at first became much better. But the supply of penicillin ran out after four days, so he died. Still, scientists immediately realized the drug's benefits.

By the end of World War II (post–1945), scientists learned how to produce penicillin in larger quantities. The process involved growing it in deep **fermentation** tanks and then separating out and purifying the penicillin. After that, people stopped routinely dying from pinpricks and shaving nicks and rose thorn scratches.

Alexander Fleming

MOLDY BUT GOODIE

Mold is a form of fungus, and a mold called penicillium naturally produces the antibiotic penicillin. The penicillin you may have taken when you were sick is a purified version of the substance. But you can grow a form of the mold.

Do not use this mold as a medicine!

Also note: This experiment takes three to five days.

YOU'LL NEED:

- **A sealable plastic sandwich bag**
- **A slice of bread—the bakery kind is best to avoid anti-fungal ingredients.**
- **Eyedropper or mist spray bottle**
- **A pen or pencil**
- **A notebook**

FOLLOW THESE STEPS:

1. Put about ten drops of water into the bag, or spritz inside the bag a few times with your spray bottle.

2. Put the bread into the bag.

3. Close and seal the bag.

4. Keep the bread and bag in a warm, dark place for three to five days and watch what happens over time.

5. Record your observations in your notebook.

What's Going On?

Did your bread become moldy? If so, it's likely that mold **spores** traveled through the air and landed on the soggy bread, where they grew. (For more about spores, see page 155.) Note that not all mold is penicillium. Penicillium starts out gray or white, and gradually turns blue-green, sometimes with a white outer ring.

SECRET TO A LONG LIFE

Could a living thing survive for a hundred years or more?

OLD

Plants, trees, fungal species, and bacteria have been known to survive for decades, centuries, and in some cases for thousands of years. And even some animals, such as sponges, coral, sharks, clams, sea turtles, and yes, humans, can live longer than a century.

Sea Turtle: 80-150 years

Greenland shark: 300-500 years

Giant Sequoia: up to 3,000 years

IMMORTAL

Other species of animals appear not to age at all. They're not *technically* immortal (which would mean they literally live forever). They can die from diseases or be killed in other ways, but their bodies don't appear to deteriorate with time. Animals that experience some version of biological immortality include certain species of clam, jellyfish, and yes, those ever-amazing planarian flatworms (see page 76).

Naked mole rat: Up to 30 years

Hydra: Up to 1,400 years

Immortal jellyfish: A really, really long time

ANCIENT

Some types of bacteria spores thought to be millions of years old have been revived and have **reproduced**. Seeds of plants that are thousands of years old have **germinated** in laboratories.

Judean date palm: 2,000-year-old seeds have sprouted

Narrow-leafed campion: 30,000-year-old seeds have sprouted

Nematode: 40,000+ years

LONG SLEEPERS

Many animals can sleep for long periods of time. They do it to conserve energy. Some of these animals **hibernate**, including woodchucks and ground squirrels. Some enter a sleep-like state for most of the winter, although unlike regular sleep, their heart rates drop considerably and their brain activity can be undetectable.

HIBERNATION

"True" hibernators are almost impossible to wake up. The animal's body temperature drops, and its breathing, **metabolism**, and heart rate lower. It's an amazing adaptation to harsh conditions.

Lemur

Bat

Hedgehog

TORPOR

Skunks, chipmunks, and hummingbirds are not "true" hibernators, in that they are light sleepers and can be awakened. (Scientists debate about whether or not bears are "true" hibernators.) But all these species enter a state of **torpor** during winter. It's a little like hibernation in that the animal's metabolism slows way down to conserve energy. The state of torpor may last from a few hours to a few weeks.

Chickadee

Bear

Hummingbird

ESTIVATION

Other animals enter a state called **estivation**. Although similar to hibernation, estivation usually happens in summer, or during hot, dry periods, rather than in winter.

Ladybug

Snail

Crocodile

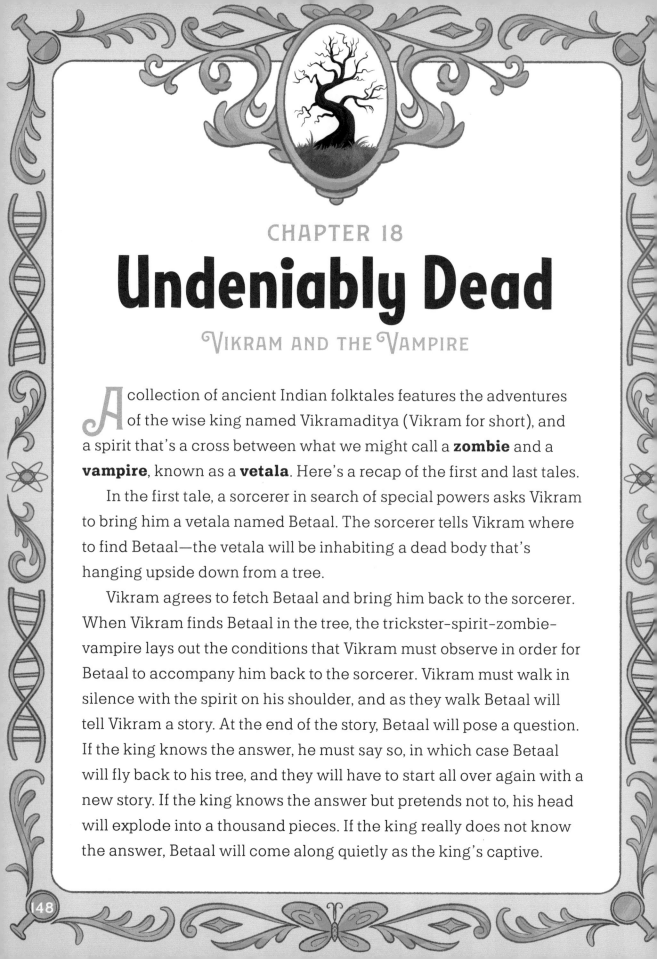

Undeniably Dead

VIKRAM AND THE VAMPIRE

A collection of ancient Indian folktales features the adventures of the wise king named Vikramaditya (Vikram for short), and a spirit that's a cross between what we might call a **zombie** and a **vampire**, known as a **vetala**. Here's a recap of the first and last tales.

In the first tale, a sorcerer in search of special powers asks Vikram to bring him a vetala named Betaal. The sorcerer tells Vikram where to find Betaal—the vetala will be inhabiting a dead body that's hanging upside down from a tree.

Vikram agrees to fetch Betaal and bring him back to the sorcerer. When Vikram finds Betaal in the tree, the trickster-spirit-zombie-vampire lays out the conditions that Vikram must observe in order for Betaal to accompany him back to the sorcerer. Vikram must walk in silence with the spirit on his shoulder, and as they walk Betaal will tell Vikram a story. At the end of the story, Betaal will pose a question. If the king knows the answer, he must say so, in which case Betaal will fly back to his tree, and they will have to start all over again with a new story. If the king knows the answer but pretends not to, his head will explode into a thousand pieces. If the king really does not know the answer, Betaal will come along quietly as the king's captive.

So twenty-four times in a row (each one its own tale), Betaal tells Vikram a story and ends it with a brain teaser, and twenty-four times in a row Vikram, who is very wise, is able to solve the riddle and answer the question correctly, so Betaal flies away and back to his tree. But finally, in the twenty-fifth tale, Betaal tells a story and Vikram is unable to answer the concluding riddle. Here's a recap of Betaal's twenty-fifth story.

A terrible war is raging in a kingdom. The king is killed, and his queen and their daughter run for their lives and take refuge in a jungle. Meanwhile, a father and his son pass by and rescue the queen and princess from their perilous predicament. In due time, the father marries the princess and the son marries the queen. Then the father and the princess have a baby boy, and the son and the queen have a baby girl.

Now comes Betaal's riddle. He asks Vikram: What is the relationship between those children?

Finally Vikram is stumped. Which means finally Vikram is able to deliver Betaal to the sorcerer.

But Betaal has taken a liking to Vikram, so he gives him some vital information before they reach the sorcerer. When Betaal had been alive, he had been a nobleman with many talents. The sorcerer had tricked him and then killed him in order to obtain his powers, and Betaal is now cursed to take the form of a vampire-zombie until a king comes along who can answer twenty-four of the riddles. And now Betaal wants revenge on the sorcerer. He tells Vikram that Vikram has been set up. He tells Vikram that the sorcerer also plans to kill Vikram in order to obtain his powers for himself and so that the sorcerer can rule the world. Betaal advises Vikram to kill the sorcerer, and explains how he can pull it off.

TALE ORIGIN

These Indian tales were written down around the eleventh century. But they're probably much older than that.

If he does, Betaal tells Vikram he will have Betaal's gratitude forever.

Vikram doesn't completely trust Betaal, but he prepares himself in case the trickster-spirit's predictions about the sorcerer prove to be true. Sure enough, the sorcerer does try to kill Vikram, but having been forewarned, Vikram outwits him and kills the sorcerer instead. Vikram becomes all-powerful, while Betaal has achieved his freedom and his revenge. Betaal promises to serve Vikram from now on.

Are zombies and vampires real?

The Scientific Scoop: The Very Hungry Caterpillars

The first zombie stories originated in Haiti, several centuries ago. In Haitian folklore, magic turned people into mindless creatures that shuffled around in a deathlike state. Later, some novels and movies appeared that depicted zombies as dead people who, through various unscientific methods, turned into the "undead" and shambled around, preying on the living and eating their brains. But in the real world, undead human zombies don't exist.

But in the *natural* world, zombies actually do exist. Certain species can parasitize other species, taking over their brains and controlling their behavior. (See more about **parasites** on pages 195-196.)

Vampires

In western mythology, vampires are undead fiends that feed off the blood of the living, using sharp fangs to puncture the necks of their victims.

You've probably heard about the vampire character known as Dracula. He was a vampire in a nineteenth-century novel, and may or may not have been based on a real person named Prince Vlad Dracula, who lived in the 1400s in what is today the country of Romania. There have been many fictional vampires before and since Dracula, in many different cultures.

Luckily, vampires don't exist in the real world, right?

Um, wrong. In the animal world, there really are bloodthirsty creatures that pierce or cut the skin of other species in order to drink their blood. This practice is known as **vampirism**. (See page 156.)

TRY THIS

MAKE ZOMBIE BRAINS

Here's a recipe for slime, which may or may not resemble gooey, drippy zombie brain-food. Whatever. This slime is really fun to play with.

NOTE Slime can be sticky and messy, so make it on a surface you can easily clean afterward.

YOU'LL NEED:

- **Mixing bowl**
- **Wooden spoon**
- **120 milliliters (1/2 cup) school glue**
- **60 milliliters (1/4 cup) hot water**
- **15 milliliters (1 tablespoon) saline or contact lens solution**
- **5 grams (1 teaspoon) baking soda**
- **Food coloring—green seems to be the preferred color for zombie brains**
- **A pen or pencil**
- **A notebook**
- **Optional: Plastic storage bag**

FOLLOW THESE STEPS:

 1. Stir together the glue and hot water in a mixing bowl. (Or use more glue and less water for a firmer texture.)

2. Add a few drops of food coloring and stir until no white streaks remain.

3. Stir in the baking soda and saline solution.

4. Knead well with your hands. If it's too sticky to work with, add an extra squirt of solution.

5. Play with your slime. Stretch it, squish it, pour it on the table. Try pulling it apart and putting it back together.

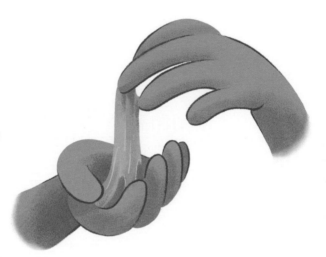

6. After you've finished cleaning up, record your observations in your notebook.

 NOTE When you are finished with your slime, dispose of it in the garbage. Don't pour it down a drain.

What's Going On?

Have you created a chemical reaction? Why or why not?

Chemicals in the contact lens solution react with the glue and produces slime's squirmy texture. Is your squishy slime a **solid** or a **liquid**? Slime is what we call a **non-Newtonian liquid**. It doesn't behave strictly like a solid or a liquid.

INVASIONS OF THE
BODY SNATCHERS
Real-life zombies and vampires

Warning: Do not read this section of the book while eating your lunch.

PHORID FLY

Phorid flies are relatively harmless to plants and to humans. Fire ants, on the other hand, cause serious agriculture problems, and their sting really, really hurts. Scientists have discovered that a great way to get rid of fire ants is to release phorid flies into an ant-infested area. The flies turn the ants into zombies. Here's how the flies do it:

The female phorid fly darts out of the air and lays an egg on a fire ant's leg. When that egg hatches, the phorid fly larva emerges and then burrows its way inside the fire ant, feasting on the all-you-can-eat ant-guts buffet. The fly larva also takes over the ant's mind, driving its still-living body around like a zombie go-cart. Eventually, the fire ant's head falls off. Over the next two weeks, the ant's dead body continues to provide baby food for the growing phorid fly larva.

JEWEL WASP

A jewel wasp can be a cockroach's worst nightmare because it can turn a cockroach into a zombie.

To zombify its victims, the jewel wasp injects a mind-altering venom into the cockroach's brain (technically the part called its **ganglia**). The venom causes the cockroach to lose all will to flee. Then the tiny wasp leads her much-larger victim by the antenna—a zombie on a leash—into her burrow, where she attaches her egg to the cockroach's leg. Upon hatching, the baby wasp larva burrows its way into the cockroach and feeds on the ready-to-eat, still-not-dead victim, sometimes for weeks, before bursting out of its now-dead carcass as a fully grown wasp, ready to start the cycle all over again.

CORDYCEPS FUNGUS

Deep in the Amazon jungle lurks a parasitic fungus called Cordyceps. If one of its tiny **spores** lands on an unlucky ant, the fungus enters the ant's body, floods its brain with chemicals, and takes control of the ant's muscles. The zombie ant staggers off to a place where conditions are perfect for the fungus to thrive. Following its fungus-master's instructions, the zombie ant opens its jaws and clamps down on the underside of a leaf, unable to let go. And just when you think it can't possibly get any worse for the ant . . . it gets worse.

Slowly, over the course of about three weeks, a horrifying-looking stem-like thing emerges from the ant's head. Then the tip of it bursts like a piñata. It showers the immediate area with more zombie-creating spores. The fungus can wipe out entire ant colonies. There are hundreds of different types of Cordyceps fungi, and each one specializes in infecting its preferred species.

NEMATOMORPHA

This charming creature, affectionately known as a hairworm, is born in the water but parasitizes insects that live on land. An unsuspecting cricket (or grasshopper, or spider) ingests the microscopic parasitic larva. Once inside the insect or spider's body, the hairworm pumps out a cocktail of chemicals that zombify its **host**. Nested inside its comfy lodgings, the hairworm grows rapidly. When it's ready to get back to the water to breed, the hairworm mind-controls its host, compelling it to stagger to the water's edge and then to take a swan dive into the water, where it drowns. The worm then emerges from the body of its drowned host and can complete its life cycle.

VAMPIRE BAT

Yes, vampire bats really exist. They live in parts of Mexico, Central America, and South America. There are lots of different kinds of vampire bats, and some of them prefer to dine on ordinary bat food like insects and fruit. But there are three species of vampire bats that feed only on the blood of other animals. Luckily human blood tends not to be their first choice—they usually prefer the blood of livestock, such as cows, pigs, and horses, as well as that of birds. They puncture their victim with their tiny sharp teeth and then lap up the blood that drips out. Their spit contains a protein called— this is not a joke—Draculin, which keeps the blood from clotting.

MOSQUITO

When someone says "mosquito," you probably don't immediately think "Ah! A classic example of a vampire!" Well, you probably should. These flying syringes cause more misery worldwide than any other insect. Only the female mosquito drinks blood. She uses the protein in blood to form her eggs. At least sixty species of mosquitoes worldwide transmit deadly diseases, including malaria, yellow fever, encephalitis, and Zika.

VAMPIRE FINCH

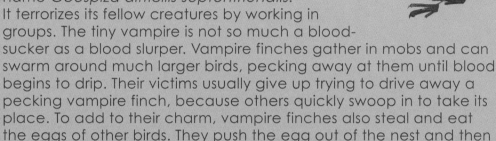

It may look small and harmless from a distance, but up close you can see its cruelly sharp beak. This species of finch, a native of the Galápagos Islands, goes by the scientific name *Geospiza difficilis septentrionalis*. It terrorizes its fellow creatures by working in groups. The tiny vampire is not so much a blood-sucker as a blood slurper. Vampire finches gather in mobs and can swarm around much larger birds, pecking away at them until blood begins to drip. Their victims usually give up trying to drive away a pecking vampire finch, because others quickly swoop in to take its place. To add to their charm, vampire finches also steal and eat the eggs of other birds. They push the egg out of the nest and then swarm around the broken egg, slurping up the yolk.

VAMPIRE MOTH

There are over seventeen different species of *Calyptra* moths, and eight of these feed on the blood of mammals. Most types of vampire moths live in areas of Asia, Africa, and Europe. This creature uses its long, piercing and sucking mouthpart (called a **proboscis**) to penetrate even the thick hides of elephants, buffalo, and other large mammals. Once through the hide, the insect drills its proboscis farther in and then, when the blood begins to flow, opens hooks to keep its slurper in place while the moth drinks its blood soup.

Vampire moths have been known to prey on humans as well. While the idea of being on the dinner menu of a vampire is deeply unpleasant, the good news is, these moths don't appear to transmit diseases.

CHAPTER 19

Poultry in a Panic

CHICKEN LITTLE

Chicken Little is out in the barnyard one day when a bird flying overhead drops an acorn. The acorn bonks Chicken Little on the head. She instantly goes into freak-out mode, convinced that the sky is falling. She dashes off to warn everyone.

First she runs into Henny Penny, and Chicken Little squawks that the sky is falling. Henny Penny believes this fake news, and joins Chicken Little in a frantic race to warn the rest of the barnyard. One by one they encounter Goosey Loosey, Turkey Lurkey, and Ducky Lucky, and all of them join Chicken Little and Henny Penny in a mad dash to warn more livestock and create mass hysteria.

TALE ORIGIN

This twentieth-century version of the story is the one that's best known to American kids—in England the tale is more commonly called "Henny Penny" or "Chicken-licken." But folktales about mass hysteria and panic can be found around the world, and date back centuries.

In the darker nineteenth-century version of this tale, the gaggle of freaked-out birds runs into Foxy Loxy. With a sly smile, Foxy invites them all into his den to "protect" them. And then he eats them all for supper. The End. Welcome to nineteenth-century bedtime stories, guaranteed to give young children nightmares!

In the more upbeat twentieth-century version of the tale, the foolish fowl beat a path all the way to the king, who calms them down. He tells them there's no need to panic, because the only thing that can fall from the sky is rain. And there that version of the story ends.

The moral of this tale? Don't jump to conclusions, and don't believe everything you hear, because your life may or may not depend upon it.

**What do we actually mean by "the sky,"
and could it really fall?**

The Scientific Scoop: What's Up?

The scientific name for "the sky" is the **atmosphere.** On Earth, the atmosphere is a blanket of gases held in place by gravity. So the short answer to our scientific question is no, this cloud of gases, mostly nitrogen, oxygen, argon, water vapor, and carbon dioxide, cannot fall and hit you on the head. That said, Chicken Little's panic is not completely irrational, because in fact there's a long history of ominous things that have fallen out of the sky. After all, hardhats were invented for a very good reason. (See page 166.)

And despite what the king says to reassure Chicken Little and her friends, rain is not the only form of **precipitation** that falls from the sky. There are many other types, and sometimes when they fall out of the sky and hit you on the head, it can hurt. Take sleet, for example. Sleet forms when raindrops freeze on their way down, and it can be unpleasant to walk outside when it's sleeting. Sleet can also cause hazardous driving conditions. Then there's hail. Hailstones could certainly fall and bonk you on the head. Hail forms when sleet starts to fall and then gets carried back up by rising air currents. The sleet pellets get covered by more rainwater, freeze again, and fall again. If this cycle repeats itself enough, the hailstones can get very large.

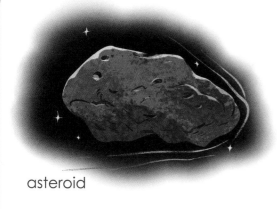

asteroid

Let's look more closely at one other type of **celestial body** that could theoretically fall from the sky and hit you on the head: it's known as a **meteorite**.

Far beyond Earth's atmosphere there are relatively small (in the sense that they're not planet-sized), rocky objects that revolve around the sun. They're called **asteroids**.

Sometimes an asteroid enters Earth's atmosphere. When it does, it's known as a **meteor**. Most of the "shooting stars" you see in the night sky are not stars at all. They're actually meteors that burn up before they can fall to the ground and do any damage. But occasionally a piece of the meteor actually strikes the earth's surface. That's called a meteorite. (Note that geologists use the word "asteroid" for any meteor that is greater than 1 kilometer [about 6/10 mile] in diameter.)

Asteroids also strike other planets and moons in our galaxy. Venus is the only planet that spins backward, relative to other planets. Many scientists believe that when the planets were just being formed, Venus got hit by an asteroid that was so huge it reversed the planet's spin. Scientists also believe that another huge asteroid hit Earth around the same time, breaking off a chunk of the planet in the process. According to this theory, the chunk that broke away eventually became our moon. The pocked surface of our moon is the result of many asteroid strikes. Because the moon has no atmosphere, wind, or water, its cratered surface has remained relatively unchanged.

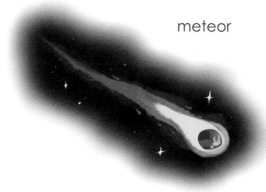

meteor

TRY THIS

Crater Maker

Create your own craters.

 NOTE This experiment is messy! It's best to do it outside.

YOU'LL NEED:

- **An oblong cake pan**
- **200–600 grams (3–5 cups) of flour, depending on the size of your cake pan**
- **Small objects to drop such as coins, marbles, or golf balls**
- **A sturdy chair or step stool**
- **A ruler**
- **A pen or pencil**
- **A notebook**
- **Optional: A box of chocolate pudding mix, for color contrast**

FOLLOW THESE STEPS:

1. Add flour to the pan so that there is about a 2 1/2 centimeter (1 inch) layer. Gently shake the pan to make sure the flour is even. You can sprinkle a thin layer of pudding mix over the flour to make your craters easier to see.

2. Stand over the pan, holding one of your objects at chin level.

3. Drop the object into the flour.

4. Measure the size of the crater from one edge to the other.

5. In your notebook, record the measurement and the height from which you dropped your "meteorite."

6. Shake the pan gently to smooth the flour, and try dropping an object that is slightly larger.

7. Record the measurements of the new crater.

8. Try dropping the same object from a higher height (such as above your head, and then while standing on the step stool or chair).

9. Measure and record the size of each "crater," what size object made them, and from how high above the pan the object was dropped.

10. Record your observations in your notebook.

What's Going On?

How do craters formed by small things compare to those formed by larger ones? Did the height at which the object was dropped affect the size?

Small objects can make relatively large craters if they hit the surface at a high speed. The object's **kinetic energy** creates a crater upon impact.

CHALLENGE! Cretaceous Crater

If a large meteorite were to hit Earth, what effect would it have?

Approximately 66 million years ago, a huge object from outer space—approximately 12 kilometers (7 1/2 miles) wide—slammed into the earth at what is now the Yucatán peninsula of Mexico. It formed a crater 150 kilometers (93 miles) wide. It hit the earth at a speed that's hard for us to imagine—probably 16 to 32 kilometers per second (10 to 20 miles per second). Today the impact site is known as the **Chicxulub crater**, after the nearby town of that name. Scientists are still debating what the object was—it might have been an enormous meteorite, or possibly a **comet** (a cosmic snowball made of frozen gases, rock, and dust). You usually see it referred to as the "Chicxulub asteroid," also called an **impactor**.

In physics terms, the kinetic energy in the zooming asteroid caused an enormous explosion upon impact. A huge cloud of debris shot into the atmosphere, while the heat produced by the impact caused a thermonuclear explosion—basically a big bomb. According to a widely accepted theory, the result of the impact wiped out about three-quarters of life on Earth, including the dinosaurs.

Scientists believe that the dust from the impact—a colossal cloud of vaporized soot and sulfur—would have blotted out most of the sunlight for years, cooling off the planet and killing huge amounts of plant life. The Chicxulub crater is egg-shaped rather than round. What does that tell you about the angle at which the impactor struck? How might that have affected the resulting path of the dust?

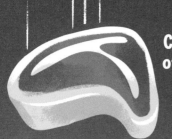

Chunks of meat

Tadpoles

Raw chicken

On a cloudless March day in 1876, residents from the town of Olympia Springs, Kentucky, discovered it was snowing chunks of meat. After scientists examined the bits of meat all over the place, they determined that the most likely explanation was **projectile vomit**. Apparently, a large group of vultures ate the meat and then regurgitated it as they flew overhead. Today, that odd day is known as the Great Kentucky Meat Shower.

In 2009, residents of a small town in Japan found themselves caught in the midst of a dead tadpole downpour. Scientists decided the strange precipitation was the result of strong winds that sucked tadpole-filled waterspouts into the sky during severe storms.

In Virginia in 2012, pieces of raw poultry fell from the sky, landing on the heads of several astonished people. The likely source? Buzzards. The buzzards had eaten some dead chickens from a nearby poultry-processing plant and had then upchucked some of their meal onto the ground below.

HEADS UP

THINGS THAT HAVE FALLEN FROM THE SKY

Chicken Little may have had a legitimate reason to be concerned after getting bonked on the head by that acorn. Should we be concerned, too?

Of course there are the obvious things that can fall out of the sky: different types of **precipitation** (thanks, king!). And let's not count human-made falling objects, such as space junk, construction materials, and falling pianos. What are some other things that have fallen out of the sky?

Live Fish

A Shark

Spiders

In 2004, residents of a town in Australia's Northern Territory noticed that hundreds of small white fish—called spangled perch—were raining down from the sky. Most of the fish were alive before they hit the ground. Meteorologists suggested that a tornado might have sucked up water and fish from a river several hundred kilometers away. The fish had been flash-frozen and carried along in the updraft high above the ground, before being dropped to Earth.

In 2013, thousands of live spiders fell from the sky in the small Brazilian town called Santo Antonio da Platina. This particular spider species, *Anelosimus eximius*, is a very social one. The spiders hang out together in big groups, and can build webs 20 meters (65 feet) high. If a gust of wind detaches the webs from their anchor points, the spiders catch a ride on the breeze and get carried along before raining down.

In 2012, a live, 1/2-meter (2-foot) leopard shark fell out of the sky and landed, flopping around, on the twelfth tee of a California golf course. No one was quite sure how it got there, but a leading theory was that the shark had been plucked from the water by a large bird and then dropped. The shark was released back into the ocean.

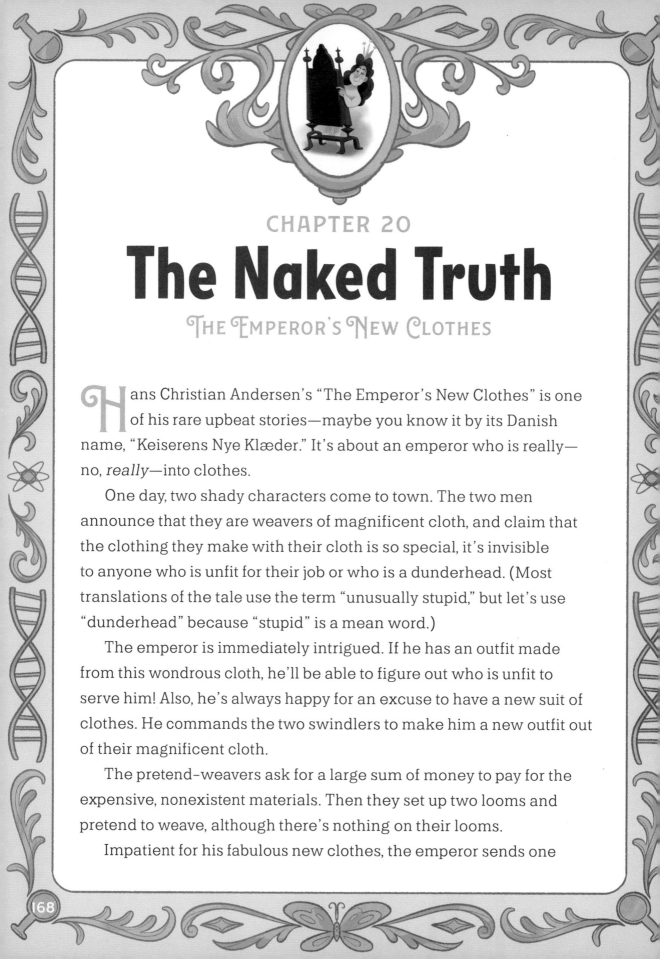

The Naked Truth

THE EMPEROR'S NEW CLOTHES

Hans Christian Andersen's "The Emperor's New Clothes" is one of his rare upbeat stories—maybe you know it by its Danish name, "Keiserens Nye Klæder." It's about an emperor who is really— no, *really*—into clothes.

One day, two shady characters come to town. The two men announce that they are weavers of magnificent cloth, and claim that the clothing they make with their cloth is so special, it's invisible to anyone who is unfit for their job or who is a dunderhead. (Most translations of the tale use the term "unusually stupid," but let's use "dunderhead" because "stupid" is a mean word.)

The emperor is immediately intrigued. If he has an outfit made from this wondrous cloth, he'll be able to figure out who is unfit to serve him! Also, he's always happy for an excuse to have a new suit of clothes. He commands the two swindlers to make him a new outfit out of their magnificent cloth.

The pretend-weavers ask for a large sum of money to pay for the expensive, nonexistent materials. Then they set up two looms and pretend to weave, although there's nothing on their looms.

Impatient for his fabulous new clothes, the emperor sends one of his trusted advisors to check on the progress of the two men. The trusted advisor goes to see them. To his dismay, he can't see any cloth, thread, or buttons at all, but he pretends to the two swindlers that he can. Secretly the man worries: Is he unfit to be a trusted advisor to the emperor? Is he a dunderhead? The swindlers ask him what he thinks. He tells them their work is enchanting, and that the patterns and colors are exquisite. The embarrassed advisor goes back to the emperor and delivers a glowing progress report.

The emperor sends another trusted advisor, who also sees nothing. But he, too, fears he is unfit for his position, or possibly a dunderhead, so he reports that the weavers are doing a splendid job.

The emperor goes to see for himself. He arrives with a big group of noblemen from his court, including the two trusted advisors. The two advisors rave about how magnificent the fabric is. None of them, including the emperor himself, sees anything, of course. Now the emperor privately wonders if he is unfit to be emperor. But he loudly exclaims how pretty the cloth is, and the rest of the group agrees.

The swindlers stay up all night pretending they're putting the finishing touches on the new suit of clothes for the emperor. The emperor plans to wear his new suit in a procession the next day. The following morning the swindlers help the emperor into his new outfit.

They pretend to button invisible buttons and fuss around adjusting invisible pleats and ruffles. At last they "finish," and tell him how fabulous he looks, but in fact he's just standing there naked. All his advisors gush about how great he looks in his splendid new outfit because everyone is afraid of what others will think of them. Off he goes to march naked in his procession, and the swindlers skip town.

All the townspeople along the route admire the emperor's new clothes. No one wants to admit that they can't see anything for fear of being called a dunderhead.

Then a little kid in the crowd yells out: "But he hasn't got anything on!" Although the child's embarrassed father shushes her, the child's words get passed from person to person along the procession route.

The townspeople all realize the child has said what everyone else was afraid to say. The emperor is wearing no clothes at all.

The emperor has also heard what the kid yelled, and even he privately suspects the kid is right. But he decides the procession must go on, and continues on his way, as his royal advisors keep pretending to hold up a long train that isn't there at all.

The story highlights an interesting issue. Sometimes a large group of people rapidly and dramatically moves from a seemingly unshakable belief to changing their minds or adopting a new behavior. Social scientists call this rapid change a "tipping point." Sometimes tipping points happen when a few key influencers decide something and talk about it publicly, and the crowd follows their lead. Sometimes all it takes is one person with the courage to say what others are afraid to—known as a whistleblower—to change public opinion.

Tale Origin

Hans Christian Andersen published this story in 1837, although he wrote that he had based his version on an old Spanish folktale with a similar plotline.

EXPLORE the SCIENCE

How could an observer think they see something that's not really there?

The Scientific Scoop: Believe It or Not

In the story, people did not see the emperor's clothes, nor did they believe they did—they simply pretended to because they were afraid of what other people might think of them. But is there a situation where you really *do* think you see something that isn't actually there?

Creating an illusion (something that looks convincing but is not real) has been a goal for artists ever since they began trying to represent three-dimensional objects on a two-dimensional surface. This type of illusion is known as "perspective," and it's the art of creating the appearance of space and depth on a flat surface.

perspective

Your eyes take instructions from your brain. Seeing is a combination of your eyes capturing a picture and your brain interpreting what the picture means. An **optical illusion** is an image that tricks our brain into seeing something that is not real. Our brain tries to interpret what our eyes have seen.

A successful optical illusion gets you to "see" something that isn't there. Or you may *not* see something that *is* there. Or different people may look at the same image and see different things.

TRICK YOUR EYES

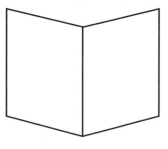

Look at this image that represents a folded-up piece of paper. Is it pointed away from you or open in front of you?

What do you see when you look at this picture? Is it two faces in profile, or a goblet?

Visual "confusion" can also result from an area known as your **blind spot**. That's a part of your eye that can't see. It's the place where your nerves come together and exit your eye to go to your brain.

Here's how to find your blind spot:

Close your right eye. With your left eye open, hold the book at arm's length from your face while staring at the ✚. Now move the book slowly toward you.

Did the circle disappear? If so, you've located your blind spot.

With your right eye closed, stare at the dot below with your left eye. Slowly move the book toward your face.

What happened? Your brain filled in missing information and connected the broken lines.

How can we trick the eye with colors or shapes? Try this:

Stare at the image to the right for twenty seconds. Then immediately look at a white wall.

Now check out his checks. Do your eyes "see" black dots at the intersection of the white lines? Try drawing and coloring a new suit for the emperor using some form of contrasting colors or shapes that "trick" your eyes.

CHAPTER 21

Just Ducky
THE UGLY DUCKLING

A mother duck sits on her eggs, impatient for her ducklings to be born. At last, all the eggs hatch—except for one. Finally the last egg hatches, but the creature that emerges is not like the other ducklings. Although the mother duck loves all her babies equally, the one that hatched last is much bigger than his siblings and is nowhere near as cute. He becomes known as the Ugly Duckling, and he endures much teasing and bullying from the other animals in the barnyard. Eventually, the Ugly Duckling runs away and tries to blend in with some wild ducks and geese, but they aren't very nice to him, either.

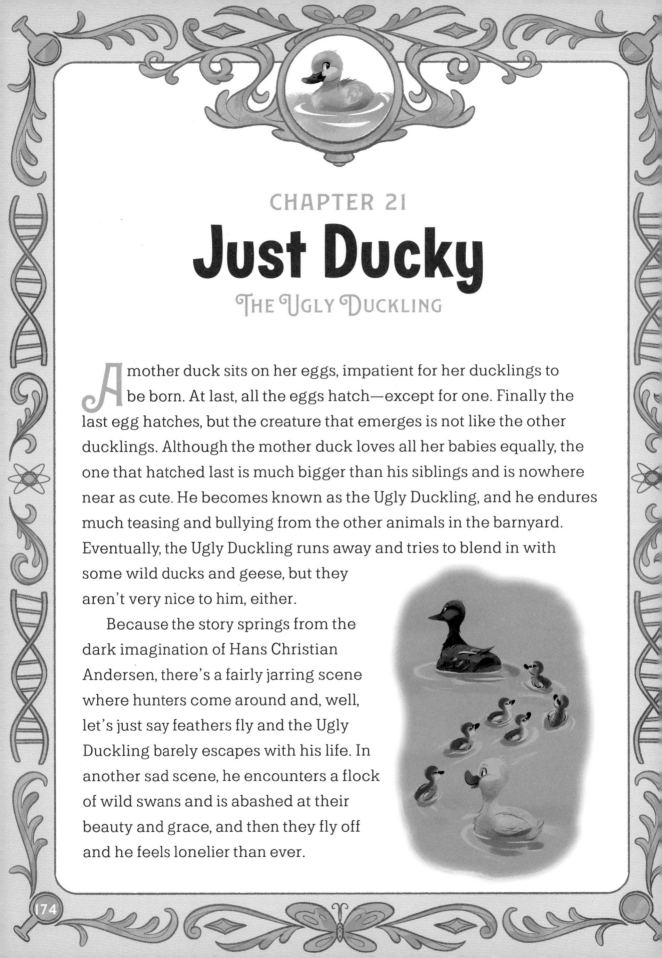

Because the story springs from the dark imagination of Hans Christian Andersen, there's a fairly jarring scene where hunters come around and, well, let's just say feathers fly and the Ugly Duckling barely escapes with his life. In another sad scene, he encounters a flock of wild swans and is abashed at their beauty and grace, and then they fly off and he feels lonelier than ever.

The tale grows ever bleaker. Winter arrives, and the Ugly Duckling gets hungrier and hungrier and colder and colder until at last his little patch of pond freezes completely and he gets stuck fast in the ice. A farmer rescues him and brings him home, but the Ugly Duckling gets frightened by the houseful of noisy kids and flees again. He barely survives the rest of the winter.

TALE ORIGIN

This tale was published by Hans Christian Andersen in 1843, and seems to have been an original tale. He later claimed that the tale was something of an autobiography.

At last spring arrives, and the flock of beautiful swans returns.

The Ugly Duckling fearfully approaches them. To his surprise, they welcome him into the group. Then he catches a reflection of himself in the water and realizes he's not a duckling at all—he's a swan just like them. By now he's fully grown, and he spreads his immense wings and flies off with his newfound relatives.

Yes, some troubling threads run through this tale: Characters judge one another (and themselves) based solely on their looks. Characters welcome strangers into their group, but only if the strangers look just like them. But let's focus on a scientific part of the story: Are there animals in the natural world that raise the young of other species?

Under what circumstances might one species adopt the young of another species?

The Scientific Scoop: Birds of a Feather

There are many examples in the bird world that mirror what happened in the "Ugly Duckling" story. Birds of certain species do sometimes lay their eggs in the nests of other bird species.

catbird

brown-headed cowbird

The brown-headed cowbird, a member of the blackbird family, is one of North America's most notorious party crashers. Cowbirds are sometimes called "brood parasites," because cowbirds never raise their own babies. Instead, these deadbeat birds lay their eggs in other birds' nests, and let foster bird parents raise their babies. Sometimes, when invading another bird's nest, the mother cowbird chucks out an existing egg before laying her own in its place. Usually the foster bird parents don't notice the cowbird's bigger, different-colored egg in their nest. The cowbird baby outcompetes the smaller, legitimate offspring for food.

But some would-be foster bird parents stand up to these cowbird bullies. For instance, catbirds don't play. Unlike other species of birds, the mother catbird recognizes the first egg in her nest as her own. If any eggs after the first are imposter eggs, she tosses them out of her nest.

Seabirds, storks, and herons often swap families in a different way. The young chicks of many of these species sometimes abandon the nests where they were hatched to seek out foster homes in other bird species' nests. This cradle jumping can be prompted by a strong survival instinct. If the chick's nest is overcrowded, the chick finds a less crowded nest, instinctually understanding that there might be more food in their adopted parents' nest.

In the 1930s, a scientist named Konrad Lorenz famously demonstrated that many species of baby birds, upon emerging from their eggs, become attached to the first moving object they see. He called this process **imprinting**. Lorenz made sure that he was the first moving thing seen by a group of baby geese upon hatching. The goslings imprinted on him and treated him like their parent.

In other cases, baby ducks have followed behind a white ball at a baseball game—the first moving thing they saw upon hatching.

We can't know from Hans Christian Andersen's story how the swan egg got into the mother duck's nest. But from a scientific standpoint, the story works as a convenient example of imprinting.

Imprinting occurs in certain species that are able to walk independently within a few minutes or hours of birth—like baby ducks, geese, and other baby birds. Because these baby birds are able to walk around, it's important that they not stray too far from their mother, so imprinting ensures that babies form an attachment and follow their mother. Imprinting in its strictest sense is limited to birds, but a form of it does happen with other animals. It's rare in mammals, because most species are born relatively helpless and take a lot more time to be able to move around independently. But a form of imprinting can happen in certain mammal species.

There's a major downside to baby animals that imprint on something other than their mother. Young animals that imprint on humans or other objects may have trouble socializing with their own species, or surviving on their own in the wild.

Mama!

Who's Your Mama?

Baby geese—called goslings—followed scientist Konrad Lorenz all around his Austrian estate.

Back in the 1950s, researchers succeeded in getting some young Canada geese and ducklings to imprint on a small green box that contained an alarm clock. Others followed a football.

Bottlenose dolphins are highly social animals. Baby dolphins, called calves, learn to recognize their mother's whistle before they're born. If the young calf gets separated from its mother, it follows the sound of her whistles.

Baby condor chicks born in captivity have been raised by human caregivers who wear a hand puppet that looks like a mother condor, in the hope that the condors will be able to survive in the wilderness as adults.

Researchers in China dress up in full-body panda suits to try to keep the cubs from preferring humans to pandas. Because pandas are an endangered species, scientists hope that the baby pandas born in captivity can be reintroduced back into the wild.

CHALLENGE!

Working Out Wingspans

We're not sure what species of swan the "Ugly Duckling" turned out to be. The trumpeter swan is the largest swan species. It can have a wingspan of 2 1/2 meters (8 feet). Here's a way to visualize that wingspan, as well as your own.

> **YOU'LL NEED:**
>
> - **A roll of wrapping paper, Kraft paper, or paper towels**
> - **A tape measure or yardstick**
> - **Scissors**
> - **Tape**
> - **A partner**
> - **A pen or pencil**
> - **A notebook**
> - **Optional: Colored pencils or markers**

FOLLOW THESE STEPS:

1. In an area with plenty of space, unroll the paper and measure 2 1/2 meters (8 feet). If you have sturdy paper, you can use your scissors to "scallop" the bottom edges of your wings so they look like feathers, and use your pencils or markers to color them in.

2. Tape the wings to a long wall.

3. Start determining your own wingspan.

4. Unroll the paper and lie down on it with your arms spread wide from side to side.

5. Have your partner mark from the tip of the longest finger on one hand to the tip of the longest finger on the other.

6. Measure and record your wingspan.

7. Cut it out and mount it on the wall.

8. Record your observations in your notebook.

GO FURTHER:

Here are some wingspans of other species of flying animals. You can measure and cut those out, too. (For very long wingspans, you can use string or ribbon.)

QUETZALCOATLUS NORTHROPI (a prehistoric winged reptile, also known as a pterosaur)	11 METERS (36 FEET)
PTERANODON (another type of pterosaur)	UP TO 6 METERS (20 FEET)

PELAGORNIS CHILENSIS
(a prehistoric bird)

**UP TO 5.1 METERS
(17 FEET)**

WANDERING ALBATROSS
(Diomedea exulans)

3.65 METERS (12 FEET)

CALIFORNIA CONDOR
(Gymnogyps californianus)

2.74 METERS (9 FEET)

BALD EAGLE
(Haliaeetus leucocephalus)

2.3 METERS (7.5 FEET)

GREAT BLUE HERON
(Ardea herodias)

1.8 METERS (6 FEET)

WHITE WITCH MOTH
(Thysania agrippina)

**28 CENTIMETERS
(11 INCHES)**

BEE HUMMINGBIRD
(Mellisuga helenae)

**6.5 CENTIMETERS
(2.6 INCHES)**

TINKERBELL FAIRYFLY
(Kikiki huna)

**16 MICROMETERS
(.006 INCHES)**

CHAPTER 22

Banding Together

The Bremen Town Musicians

An aging donkey grows concerned that his master no longer finds him useful and might be planning to get rid of him. So he decides to hit the road and head to the town of Bremen. The donkey is a musician, and he wants to start his own band.

Along the route, the donkey meets a dog. The dog is also getting older, and can no longer run fast enough to hunt for her master. She tells the donkey that she fears her master is going to get rid of her, so she has decided to run away. The donkey invites the dog to come along to Bremen with him. They can form the band together.

Next the donkey and the dog meet a cat, who has run away from her mistress because the cat can no longer catch mice the way she used to. The cat fears her mistress wants to get rid of her. The two invite her to join their band, and off they all go.

As they pass by a farm, they hear a rooster crowing. The three musicians compliment the rooster on his strong voice. He tells them that in spite of his fine ability to crow, he's just learned that his human farm family is expecting lots of company soon and he fears he's destined for the soup pot. The other three invite him to join them, and this bodacious band boogies on toward Bremen.

The town of Bremen is still more than a day's journey away, and when it grows dark, the weary travelers find themselves in a forest. The donkey, dog, and cat lie down beneath a tree for the night. But the rooster flies to the topmost branch and has a look around. In the distance he sees a light, and tells his companions there must be a house not far into the forest. All of them are hungry, so they head to the house to ask whoever lives there for food and shelter.

The donkey looks through the window of the big house and reports that it's occupied by a band of thieves. He also reports that

they've got a table laden down with food. The four musicians would really like a hot meal and a warm bed, so they come up with a plan to scare away the thieves. The donkey puts his front hooves on the window ledge. The dog jumps onto the donkey's back, the cat climbs onto the dog's shoulders, and the rooster perches on the cat's head. Then they all start braying, barking, meowing, and cock-a-doodle-doo-ing. To add a bit of **percussion** to their music, the donkey bashes his hooves through the window, shattering the glass. The robbers are so terrified by all this noise that they race out the door and run away.

The animals enter the house and enjoy a satisfying meal. Then they blow out the candles and settle down to sleep. Around midnight, the robbers sneak back to the house and, upon finding it dark, decide they'd been silly to be so frightened. One of them creeps inside to see if anyone is still in the house. Upon lighting a candle, he is startled by a pair of glowing eyes. That would be the cat. She jumps at him, spitting and scratching. In his terror, the robber accidentally lets the candle go out. As he blunders around in the dark, he steps on the tail of the dog, who leaps up, snarling and biting. The robber tears out of the door, and in the yard the donkey gives him a kick while the rooster starts loudly cock-a-doodle-doo-ing.

Back with his fellow robbers, the terrified man exclaims, "Ah, there is a horrible witch sitting in the house, who spat on me and scratched my face with her long claws. And by the door stands a man with a knife, who stabbed me in the leg. And in the yard there lies a monster, who beat me with a wooden club. And above, upon the roof, sits the judge, who called out, 'bring the rogue here to me.' So I got away as well as I could."

So the robbers never come back, and the four musicians live happily ever after together.

ᴛᴀʟᴇ ᴏʀɪɢɪɴ

Variations of this tale—where animals partner up with one another to live together and frighten off dangerous intruders—have been part of European, American, and South African oral traditions for centuries.
This version is from the Brothers Grimm.

What is the difference between noise and music?

The Scientific Scoop: Fine Tune

Many things in our world can produce sounds. In the simplest scientific sense, sound is made by matter (a solid, liquid, or gas) that vibrates. (See a more in-depth discussion of oscillation and frequency on page 28.) Animals, moving air, water, and even plants can produce sounds. Humans can as well. In addition to using our voices, humans have invented musical instruments to create sounds. But when does a sound "sound like" a noise, and when does it sound like music?

Defining what we mean by noise is relatively easy. Defining what we mean by music? Not so easy.

When voices, instruments, or, really, anything that creates sound waves produce random combinations of sounds that are loud and annoying, that's noise. (Note that "annoying" is not a scientific term, and can be a matter of opinion, but let's say "annoying to nearly everyone who hears it.")

Music, in its most basic sense, is any combination of sounds played for a certain **duration**. Taking it a step further, we can say that when voices or instruments create combinations of notes (pitches) that are played for a certain duration, that's music. We might also say that to be defined as music, notes must form a pleasing melody, but the word "pleasing" is no more scientific than "annoying," and is very much a matter of opinion. What sounds pleasing to one person's ear may sound annoying to another's.

A **rhythm** is a regular, repeated pattern of movements or sounds. Just by breathing or walking or having a heartbeat, we produce rhythms naturally. Think about what other rhythms, whether natural or human-made, you hear in your daily life.

So we can define music as a combination of pitches played in a particular rhythm and sequence that can be pleasing to the ear. Let's do some listening.

HEAR! HEAR!

Before hearing aids, many people who were hard of hearing used ear trumpets, which were funnel-shaped gizmos that collected sound waves and channeled them toward the person's ear.
Now make your own!

 NOTE Be careful not to poke yourself inside your ear!

YOU'LL NEED:

- **Two pieces of construction paper or other sturdy paper**
- **Tape**
- **A pen or pencil**
- **A notebook**

FOLLOW THESE STEPS:

1. Roll each piece of paper into a cone with a dime-sized opening at the small end. (It's easier to roll if you hold the paper horizontally.)

2. Tape both of the cones securely.

3. Hold the small ends to your ears and listen carefully.

4. Find rhythmic sounds to listen to (such as a friend's heart beating, your cat's purring, a faucet dripping, a car's blinker or windshield wipers, or a clock ticking).

5. Experiment with your ear trumpets by turning them around, so the small hole points out and the large hole is around your ear. How does this affect the sounds you hear?

6. Record your observations in your notebook.

What's Going On?

How do your ear trumpets affect the way you hear sounds?

As an object emits a sound, the sound waves spread out as they travel away from it. Your ear trumpets help amplify sounds that you hear. They capture and gather in sound waves and carry them into your inner ear, somewhat like a funnel, so your brain can analyze what you're hearing. It's the same reason your friend might cup her hands around her mouth as she whispers into your ear. Her cupped hands gather and contain the sound waves and funnel them toward your eardrum.

Think about different animals and the shapes of their ears. How do you think ear shape and size affect an animal's hearing?

CHALLENGE! ▸ Let's Make Some Noise

Here's a way to demonstrate oscillation and amplification.

YOU'LL NEED:

- **A large, disposable plastic cup**
- **Scissors**
- **A 30-centimeter (1-foot) length of heavy string, like cotton twine, as well as other types of cords, such as rubber-coated speaker wire, yarn, or dental floss**
- **A toothpick, binder clip, or paperclip**
- **A damp paper towel**
- **A pen or pencil**
- **A notebook**

 NOTE Parts of this activity will require an adult's help.

FOLLOW THESE STEPS:

1. With an adult's help, carefully poke a small hole in the bottom of the cup with the scissors.

2. Push one end of the string up and through the bottom hole, into the cup, and tie that around the middle of the toothpick or clip. Then pull the string taut, so the toothpick or clip sits at the bottom of the cup (you may need to break off a part of the toothpick so it fits).

3. Wrap the damp paper towel around the string, just below the cup.

4. Pull down on the paper towel, squeezing firmly. You might need to do it a few times to get a good sound going.

5. Record your observations in your notebook.

What's Going On?

What sort of sound did you produce? That sound is produced by **friction**, which sets off vibrations. As you pull the wet towel down the string, the string's **tension** changes, which causes the bottom of the cup to oscillate, and produces a sound. The sound waves move up the sides of the cup, focusing and amplifying them, so the cup acts just like a megaphone.

Experiment with different sized cups and different kinds of string.

Trumpeters, Woofers, and Tweeters

We measure frequency in **hertz** (Hz), which is the number of sound waves per second. The lowest frequency humans can typically hear is about 20 hertz.

Elephants can hear frequencies that are much lower than that. They communicate with one another using very low frequency sounds—between 1 and 20 hertz. The sounds are so low, the human ear can't hear them. These low sounds help them keep in touch with one another from as far away as 10 kilometers (about 6 miles).

The highest frequency humans can typically hear is about 20,000 hertz.

Dogs can hear frequencies that are much higher than that—as much as 23,000 to 54,000 hertz. That's why we can't hear the tweet of a dog whistle, but dogs can.

CHAPTER 23

Gulp

LITTLE RED RIDING HOOD

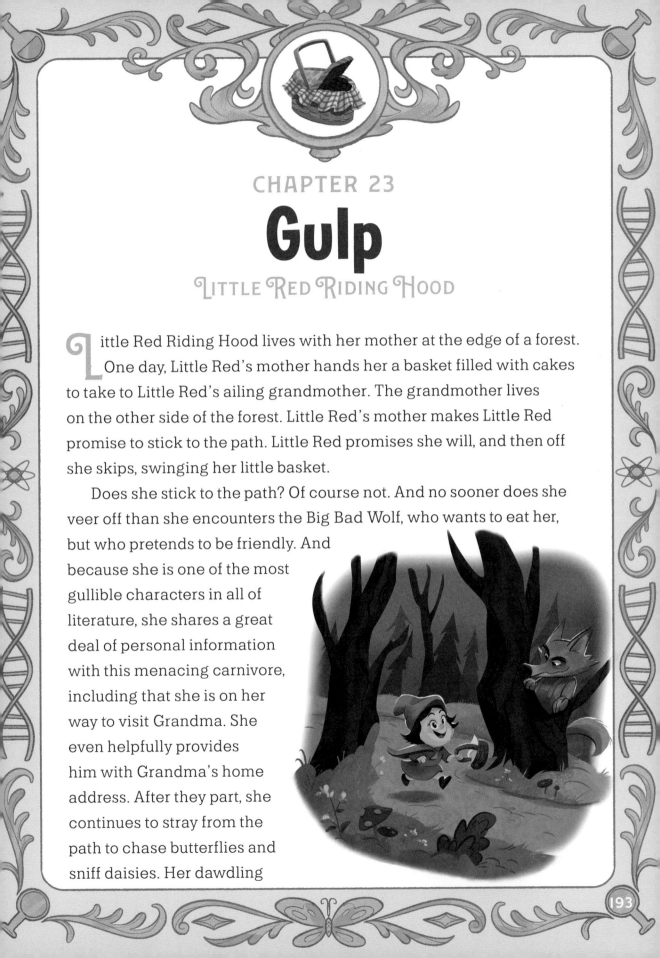

Little Red Riding Hood lives with her mother at the edge of a forest. One day, Little Red's mother hands her a basket filled with cakes to take to Little Red's ailing grandmother. The grandmother lives on the other side of the forest. Little Red's mother makes Little Red promise to stick to the path. Little Red promises she will, and then off she skips, swinging her little basket.

Does she stick to the path? Of course not. And no sooner does she veer off than she encounters the Big Bad Wolf, who wants to eat her, but who pretends to be friendly. And because she is one of the most gullible characters in all of literature, she shares a great deal of personal information with this menacing carnivore, including that she is on her way to visit Grandma. She even helpfully provides him with Grandma's home address. After they part, she continues to stray from the path to chase butterflies and sniff daisies. Her dawdling gives the wolf plenty of time to beat her to Grandma's house.

The wolf knocks on Grandma's door. He calls out to her in Little Red Riding Hood's voice. Grandma throws open the door. (Gullibility is clearly a Riding Hood family trait.) The Big Bad Wolf promptly swallows Grandma whole, puts on her nightcap and spectacles, and jumps into her bed.

When Little Red Riding Hood arrives, the wolf beckons her inside. There follows a famous exchange between Little Red Riding Hood and the wolf-in-disguise, during which they discuss how big Grandma's eyes, ears, and teeth have become. Then, the wolf swallows Little Red Riding Hood whole, too. He settles down for a nap.

A lumberjack passing by the cottage hears snoring, discovers the sleeping wolf, and somehow realizes what has happened. He cuts open the wolf with his ax. (The wolf remains asleep, which is not, to put it mildly, scientifically possible.) Little Red Riding Hood and Grandma climb out of the wolf's stomach unharmed. (More on *this* improbable plot point later.) They fill the wolf's body with stones and stitch him back up. This causes serious medical problems for the wolf. He wakes up from his nap and tries to run away, but with all those heavy stones in his belly, he dies instead.

Something is a bit off about this story, besides the improbable plot points and the fact that the girl can't tell the difference between her actual grandmother and a wolf wearing her grandmother's nightcap and spectacles. No, what's off is the *science*. How is it scientifically possible for Little Red and her grandmother to get swallowed whole, survive inside the wolf's stomach, and emerge alive?

TALE ORIGIN

This is another story that's been told around the world, and it may be thousands of years old. Very similar, ancient tales have been recorded in East Asia, Central Africa, and South Africa. This version comes from Perrault.

EXPLORE the SCIENCE

Can a living creature swallow another living creature whole, and could that swallowed creature emerge alive and well?

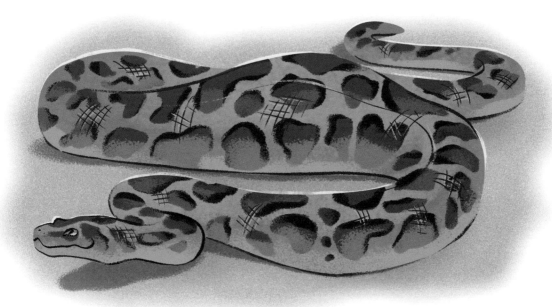

Burmese python

The Scientific Scoop: Whole Foods

There *are* animals, such as ducks and frogs, that swallow their food whole and while it's still alive. Snakes do, too. A Burmese python can swallow an entire alligator. Even some *people* swallow live animals on purpose, although people don't usually swallow their food whole. Live oysters, for instance, are a delicacy in many places. Also, *casu marzu* is a Sardinian sheep milk cheese that contains live maggots. Which people eat.

There are also tiny creatures that live and positively thrive inside larger animals' bodies. These living things are called **parasites**. Parasites (more specifically, the category known as **endoparasites**) are creatures like flatworms, tapeworms, and bot fly larvae that invade the bodies of other living things in some pretty gross ways. The living thing then becomes the parasite's unwilling **host**. Once inside the host, these parasites set up shop—sometimes for years and years. But because parasites belong to a separate and highly disgusting category of **organisms** that *want* to remain inside their hosts, we'll just disqualify them. (For more on some of these deeply unsettling parasitizing creatures, see page 148.)

So now that we know that a smaller, nonparasitic animal can be swallowed whole by a larger animal, let's turn to the second part of our question: Could that animal stay alive inside the bigger animal's digestive tract (a place without oxygen, also called an **anaerobic** environment) for a while? Can it emerge alive and well?

Why, yes. Yes, it can.

In one academic journal, scientists reported finding a live toad that had a live snake emerging from its butt. After a few minutes, the struggling toad managed to expel the entire snake. The scientists concluded that the toad had gulped down the snake quickly and with minimal jaw pressure, which enabled the snake to pass through the toad's digestive tract in a real-life *Magic School Bus*–style voyage and live to tell the tale (to other snakes).

So yes, some larger animals can swallow smaller animals whole, and yes, sometimes those small animals can survive. But unlike the Big Bad Wolf, a real-life wolf doesn't swallow its prey whole. It may wolf down its food in huge chunks, but it tears its dinner to pieces first. A whole human would be too big to fit down a wolf's esophagus and into its stomach.

TRY THIS

Swallowed Up

Here's the next best thing to swallowing a living thing whole: In this experiment you'll get a small-mouthed jar to "swallow" a whole, larger egg.

 NOTE This experiment may require an adult's help.

YOU'LL NEED:

- **A small glass jar (It should have an opening that is slightly smaller than your peeled egg.)**
- **Wooden matches**
- **Two or more peeled hard-boiled eggs (A spare helps in case one breaks.)**
- **A pen or pencil**
- **A notebook**

FOLLOW THESE STEPS:

1. With an adult's help, light two or three matches and drop them, still lit, into the jar. They'll probably go out, but that's okay.

2. Immediately put the egg on top of the jar.

3. Record your observations in your notebook.

What's Going On?

Did the egg drop all the way into the jar?

The matches caused the temperature of the air inside the jar to heat up and expand. As the air inside the jar cooled, the air pressure decreased. The greater air pressure outside the jar pushed the egg into the jar.

Swallow Tales

Meet some animals that can survive after being swallowed alive!

SNAKE VS OWL

Live Brahminy blind snakes have been found inside owls' nests high up in trees. That suggests to some scientists that the snakes had been swallowed by the owl and pooped out. Like toads, owls tend to gulp down their dinner without chewing it first.

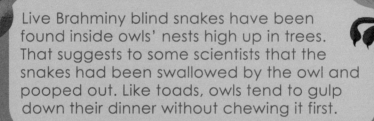

SNAIL VS DUCK

If a mallard duck swallows an aquatic snail called *Hydrobia ulvae*, the snail can often survive the five-hour journey through the duck's digestive tract. It's a handy way for the snail to hitch a ride to a new location. Another tiny snail called *Tornatellides boeningi* may actually *want* to get swallowed by its predator, a Japanese bird called *Zosterops japonicus*. Its shell acts like a little snail-shaped suit of armor, protecting the squishy animal inside from its predator's harmful digestive juices. Scientists have observed these snails giving birth to new snails . . . after getting pooped out of a bird.

NEWT VS FROG

Don't be fooled by the mild-mannered expression of the rough-skinned newt. Not only does its stinky smell help ward off predators, but its skin glands produce a potent toxin. If a frog or toad makes the poor decision to swallow this newt, the toxin quickly kills its predator. The newt then climbs right back out of its now-dead predator's stomach and continues on its merry newt way.

BOMBARDIER BEETLE VS TOAD

There are hundreds of species of these beetles, and they're known for their ability to squirt hot, toxic chemicals from their butts when threatened. Some foolish toads eat them anyway, little knowing their prey is in possession of a round-trip ticket. Once inside the toad's stomach, the beetle unleashes its horror show, causing serious gastrointestinal issues for the toad. But because toads can't vomit, the only way for the toad to upchuck the beetle is to turn its stomach inside out, a process that can take 45 minutes. The triumphant beetle emerges at last, covered in its predator's mucus-y stomach acid but otherwise alive and well.

MUSSEL VS ANEMONE

The mussel's shell acts like a super-hero's suit, repelling the stinging tentacles of a sea anemone. Should the sea anemone swallow the mussel, it can pass through the sea anemone's digestive tract and emerge just fine.

CHAPTER 24

She's the Man

HUA MULAN

In this sixth-century poem, a teenage girl named Hua Mulan learns that the emperor (known as the Khan) has called upon every family in China to send one male to fight in his army. Mulan knows that her father fully intends to serve, but he is old and sickly, and Mulan fears he will not survive the war. So she puts on men's clothes and disguises herself as a male soldier to enlist in the Khan's army in her father's place.

She carries the sword that's been passed down from her ancestors. For twelve years she fights valiantly and also maintains her disguise. We can assume she's a skilled rider and good with a bow and arrow, because she wins the admiration of her fellow soldiers and officers. In some retellings of the story, Mulan falls in love with an officer, who at first doesn't know she's a woman. When he discovers her true identity, their friendship turns to love. Eventually more soldiers come to realize she's a woman, but because she's such a good soldier they remain devoted to her. In a final battle, she shows up dressed as a woman and leads her fellow soldiers to victory.

Word of this valiant female fighter reaches the Khan. He summons her, and tells her he wants to give her a reward. She replies that her only desire is for a swift horse so that she may return home as soon as possible.

Although scholars still debate whether or not the character of Mulan is based upon a real-life person, we know for a fact that there were lots of women warriors during this period of history. They were members of nomadic tribes that traveled on horses and ponies. There were many such tribes, from China in the east, across the steppe regions of Central Asia, and all the way to areas around the Black Sea. The ancient Greeks had a name for these women warriors. They called them Amazons. A female soldier on horseback who was skilled at archery could be every bit as effective as a male soldier.

So let's talk about some of the science behind archery. Chinese and other Asian archers traditionally draw the bow with their thumbs, and wear a thumb ring for the purpose. European archers tend to pull the bow back with three fingers. You'd think an archer would close one eye to aim at a target, but—fun fact—most archers keep both eyes open. Let's investigate.

Tale Origin

Although "Hua Mulan" was written down in the sixth century, the oral story may be even older.

EXPLORE the SCIENCE

Why is it more effective to keep both eyes open when aiming at a target in the distance?

The Scientific Scoop: Look This Way

Look at an object in the distance. It should be at least 6 meters (20 feet) away from you. Close one eye and point to the object with your thumb. Open that eye and close the other. Why does the object appear to jump to the side and your finger no longer line up with the object? Now do it with both eyes open. What happens?

Having two eyes side by side on our faces gives humans **depth perception**. The process by which we see depth is called **stereopsis**. Not all animals with side-by-side eyes have depth perception. One of the clues our brains use to judge distance and depth is the very slight difference between what our left eye sees and what our right eye sees.

Our brains analyze the two pictures and then combine them to create a three-dimensional image. If you close one eye, you eliminate one of the ways your brain can judge depth.

SEE THE POINT?

Experiment with depth perception.

> **YOU'LL NEED:**
> - **Two pencils or a pair of chopsticks**
> - **A paper or plastic cup**
> - **A group of coins, paperclips, or marbles**
> - **A pen or pencil for writing**
> - **A notebook**

FOLLOW THESE STEPS:

FIRST:

1. Hold one pencil in each hand, positioned horizontally.

2. Extend your arms out fully, while still holding the pencils.

3. Close your left eye and try to touch the ends of the pencils together.

4. Open your left eye and close your right eye. Try touching the pencils again.

5. Try it with both eyes open. What happens?

6. Place the cup on the floor just in front of you.

7. Bend over very slightly. Close one eye and try to drop the coins or other small objects into the cup.

8. Repeat the process with the other eye open, then try it with both eyes open.

9. Experiment with moving the cup a little farther away from you. Try tossing the coins or other small objects into the cup with one eye closed, and then with both eyes open.

10. Record your observations in your notebook.

What's Going On?

Did you have more success dropping the object into the cup with one eye closed, or with both eyes open? Why do you think you got these results?

Because each eye sees the world from a slightly different angle, your brain combines the different pictures into one image with additional depth perception.

A Sight to Behold

Humans are not the only animals with the ability to gauge depth and distance. Monkeys, cats, horses, falcons, and toads, for instance, also have stereopsis. And guess who else? Cuttlefish! These buggy-eyed mollusks with long bodies and eight waggling arms are relatives of squid and octopuses. How do we know that cuttlefish have stereopsis? Scientists came up with a unique way to test for it. They Velcroed a pair of 3-D glasses to the cuttlefish and then projected a 3-D movie of a shrimp onto the wall of the aquarium, like an underwater drive-in movie theater. Sure enough, the bespectacled cuttlefish backed up before trying to grab the "prey" with their tentacles, showing that they were using 3-D vision to gauge the distance from the shrimp. And if you think getting a cuttlefish to sport a pair of shades is difficult, try putting them on an insect. Yes, scientists managed to stick the world's itty-bittiest 3-D glasses over the bulbous eyes of praying mantises. Turns out praying mantises also have this sophisticated skill. Stereopsis helps them snatch up their prey with their two spiky arms and then decapitate it. Who knew?

CHALLENGE! Project Projectile

If an archer on horseback shot an arrow horizontally, and at the exact same time dropped another arrow straight down, which arrow would hit the ground first?

Admittedly, shooting one arrow while dropping another at the exact same time would require quite a bit of skill, but Mulan could probably manage it.

And get this: If both arrows are released at the exact same moment—the one fired from a bow and the one dropped to the ground—then both will hit the ground at the exact same time.

As the arrow that is shot horizontally zooms forward, it is also falling toward the ground. Gravity will pull both arrows toward the earth in the same amount of time, no matter how fast one may be moving forward horizontally.

Here's a way to test the behavior of falling objects.

> **YOU'LL NEED:**
> - **Two identical coins**
> - **A narrow table or countertop located over a hard, bare floor, with space on both sides.**
> - **A pen or pencil**
> - **A notebook**
> - **Optional: A partner**

FOLLOW THESE STEPS:

 1. Place both coins side by side at the edge of the table or countertop.

2. Move to the other side of the table or countertop so that you are leaning across it.

3. Practice flicking one of the coins off the edge with your thumb and index finger. The flick motion can be tricky, so you may need to practice a few times.

4. After you've got the flick motion down, do this: At exactly the same time, flick one coin outward off the table while you nudge the other one so it drops straight down off the edge.

5. Listen carefully (or rely on your partner) to hear which coin hits the floor first.

6. Record your observations in your notebook.

What's Going On?

Did the coins hit the floor at the same time? If so, how do you think that's possible when one is moving away from the counter as it falls?

The speed at which an object falls depends only on gravity. (Air resistance does act as an opposing force, but its effects here would not affect the outcome because the coins weigh the same amount.) Because the two coins start off at the same height, the speed at which they fall is the same. It makes no difference whether one coin is moving horizontally at the same time.

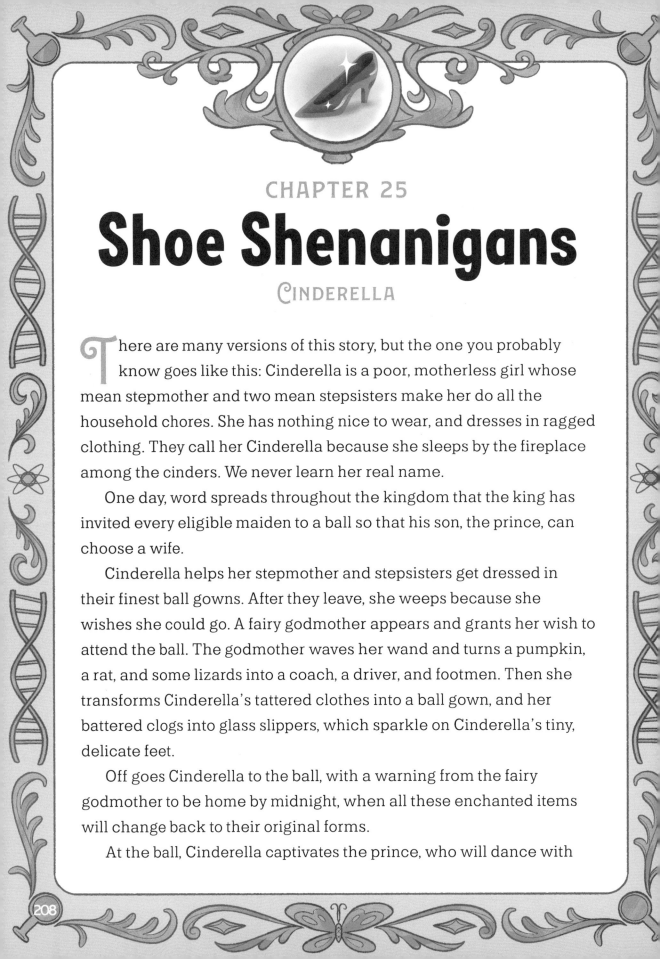

CHAPTER 25

Shoe Shenanigans

CINDERELLA

There are many versions of this story, but the one you probably know goes like this: Cinderella is a poor, motherless girl whose mean stepmother and two mean stepsisters make her do all the household chores. She has nothing nice to wear, and dresses in ragged clothing. They call her Cinderella because she sleeps by the fireplace among the cinders. We never learn her real name.

One day, word spreads throughout the kingdom that the king has invited every eligible maiden to a ball so that his son, the prince, can choose a wife.

Cinderella helps her stepmother and stepsisters get dressed in their finest ball gowns. After they leave, she weeps because she wishes she could go. A fairy godmother appears and grants her wish to attend the ball. The godmother waves her wand and turns a pumpkin, a rat, and some lizards into a coach, a driver, and footmen. Then she transforms Cinderella's tattered clothes into a ball gown, and her battered clogs into glass slippers, which sparkle on Cinderella's tiny, delicate feet.

Off goes Cinderella to the ball, with a warning from the fairy godmother to be home by midnight, when all these enchanted items will change back to their original forms.

At the ball, Cinderella captivates the prince, who will dance with no one else. Everyone wonders who this dazzling foreign princess might be. Cinderella is so busy captivating and dazzling the prince that she loses track of time, and then she hears the clock strike midnight. She manages to get away, but in her haste she loses one of her glass slippers.

The lovestruck prince finds the slipper. (It's not clear why the glass slipper remains in its enchanted state, rather than turning back into a battered old clog, but whatever.) The next day he sends his minions throughout the kingdom to try the slipper on every young woman in the land. Whoever it fits will be the one he marries. When at last the prince's servants get to Cinderella's house, each of her stepsisters tries without success to cram her foot into the tiny, delicate shoe. The wicked stepmother scoffs at the idea of Cinderella trying it, but the prince's men insist. Cinderella's tiny, delicate foot fits the shoe perfectly, which is a good thing for her, because an ill-fitting glass shoe would probably chafe a lot and create painful blisters. Let's turn our scientific gaze to those glass slippers.

TALE ORIGIN

This tale may be more than a thousand years old, and has nearly a thousand variations from all over the world. One of the earliest dates back to ninth-century China. Charles Perrault recorded "The Little Glass Slipper" in the seventeenth century.

Could glass shoes support a person's weight and withstand an evening of ballroom dancing without shattering?

The Scientific Scoop: Slipper Science

Let's establish one thing right off the bat: Slippers made out of actual glass would not survive a single waltz. We will accept this as fact because lots of engineers tell us it's so, and the mathematical formulas are extremely complicated. Engineers have access to measurable indicators such as **yield strength** (the point at which something literally gets bent out of shape) and **compressive strength** (how much resistance to breaking that a glass slipper puts up when it's on someone's foot). Based on these indicators, we can determine that while a person *might* be able to stand still in a pair of glass slippers with a modest heel, she would be unable to walk or dance in them. The bending stress from dancing would quickly cause the shoes to shatter.

What Cinderella would need is a clear, durable material that *looks* like glass—but that doesn't shatter.

Maybe her fairy godmother went with slippers made of safety glass. That's the kind of glass installed in car windshields, the windows of many public buildings, and even in some safety goggles. If safety glass gets broken, it cracks, but still maintains its original shape, more or less, rather than shattering into multiple shards and pieces. But safety glass seems unlikely. Even if the shoes broke but held together in the shape of a shoe, they'd still be hazardous to Cinderella's feet. It would be nearly impossible for her to dance around in crackly glass slippers.

How about plexiglass? That could work—although this clear acrylic substance didn't exist in Cinderella's day and would not be invented for many centuries. Plexiglass was developed in the late 1920s by various chemists at different laboratories, and was first sold in 1933. Nowadays you see it used in hockey rinks—it's that thick, see-through paneling that protects spectators from flying pucks. You can also find plexiglass in modern-day aquariums. It's strong enough to withstand a huge amount of water pressure in a giant tank. Assuming Cinderella's fairy godmother was very resourceful, plexiglass slippers would probably be Cinderella's best bet.

Modified versions of plexiglass have indeed been used as shoes, most famously back in the 1970s. During that decade, famous for its frightfully unflattering fashions, there was a (thankfully) brief fad for wearing clunky platform shoes with clear acrylic platform soles, which had removable heels. The shoe wearer could insert whatever cool trinkets they wanted into the heel. Some people filled the heels with water and live goldfish. This was a terrible idea, and a lot of innocent goldfish did not survive the night of disco dancing.

TRY THIS

CRACKING UP

Explore the yield strength and compressive strength of a different kind of fragile thing—an egg.

YOU'LL NEED:

- **Several raw, uncracked eggs**
- **Newspaper or trash bag to protect experiment area**
- **A pen or pencil**
- **A notebook**

FOLLOW THESE STEPS:

1. Take off any rings you might be wearing.

2. Hold a raw egg in your dominant hand, and wrap your fingers around it.

3. Try to break the egg by squeezing it as evenly and firmly as you can.

4. Try to break the egg in your other hand. What happens?

5. Record your observations in your notebook.

What's Going On?

What did you expect would happen? Did squeezing it with your other hand produce a different result? How about turning the egg upside down? Why is it hard to break the eggshell this way, but relatively easy to break an egg by tapping it on the edge of a bowl?

Nature has designed eggs rather well. Their curved shape ensures that a mother hen can sit on her eggs without breaking them, but her baby chicks can peck their way out of their egg when it's time to hatch. The shape of the egg distributes the force all around the curve, so the stress is spread out (See more about eggshell strength on page 137.) When you apply force with your hand wrapped around the egg, the stress gets distributed fairly evenly.

CHALLENGE! Getaway Pumpkin!

Is a round pumpkin the ideal shape for a carriage?

Think about why Cinderella's fairy godmother selected a pumpkin for the carriage rather than a differently shaped vegetable. A round pumpkin shape is not as **aerodynamic** as a vegetable with a more streamlined shape, like a zucchini. The faster the carriage goes, the more the **drag resistant forces** increase. Should the fairy godmother have opted for something longer and pointier? If not a zucchini, then perhaps a cucumber? Or more lightweight, like a bell pepper? It's possible that those vegetables were not in season at the time Cinderella went to the ball.

Experiment with creating different vegetable carriages for Cinderella. You'll need to mount your vegetable onto something that can roll, like a toy car. Consider what shape you'd like to make your carriage. Does the size or the weight of your vegetable affect how efficiently your carriage rides?

You can test the efficiency of your carriage by setting up a board on a slight incline. Measure how far your carriage rolls.

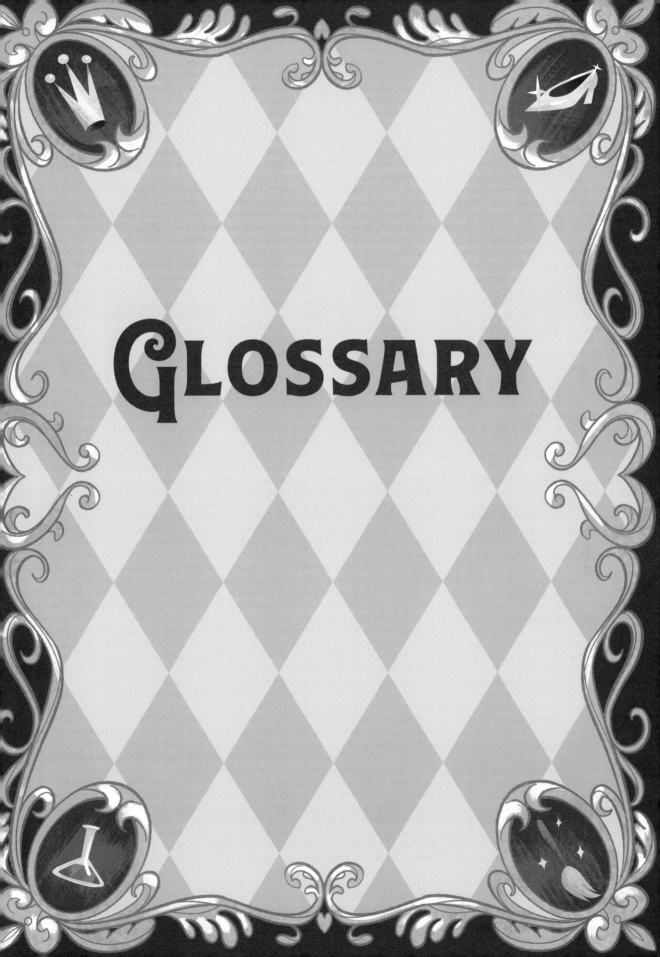

GLOSSARY

aerodynamic (air–oh–dy–NAM–ik): Having a shape that reduces drag from air moving past.

air resistance (air ree–ZI–stuhntz): A force (also known as drag) that air exerts on an object, and which tends to slow its movement. This drag force is always opposite the object's motion. The more surface area an object has, the more air particles hit it and the greater the resistance. That's why a drop of rain falls faster than does a snowflake of the same weight.

alchemy (AL–kuh–mee): An early form of chemistry, widely practiced in the Middle Ages and later. Alchemists used chemicals to try to change common metals, such as lead, into gold. Their other major goal was to create a potion that would enable a person to live forever.

alkaloid (AL–kuh–loyd): A naturally occurring chemical compound produced by living things (usually plants) that contains nitrogen. Alkaloids have been used as drugs, medicines, and poisons.

amphibian (am–FIB–ee–uhn): A class of cold–blooded animals with a backbone (vertebrates) that includes frogs, toads, newts, and salamanders. Amphibians begin life in water and later develop lungs that breathe air.

amplify (AM–pli–fy): To increase the volume or power of a sound or electric signal.

anaerobic (an–uh–ROH–bik): Living or surviving without oxygen. Also a description of a place without oxygen.

anthropology (an–thruh–POL–uh–jee): The study of people, both living and from the past.

antibiotic (an-tee-by-OT-ik): A medicine that slows down or stops bacteria from growing; a drug used to treat or prevent a disease caused by germs.

antidote (ANT-i-doht): A remedy that counteracts the effects of a poison.

arachnid (uh-RAK-nid): A member of the **arthropod** group of animals that includes eight-legged animals such as spiders, scorpions, ticks, and mites. Arachnids are close cousins of insects (which are also arthropods).

arthropod (AR-thruh-pod): a type of invertebrate animal such as an insect, spider, or crustacean.

astronomy (uh-STRAH-nuh-mee): The scientific study of stars, planets, comets, galaxies, and other parts of the universe.

atmosphere (AT-muh-sfeer): A mixture of gases that surround a planet or star; on Earth, it's also known as air.

bacteria (bak-TEER-ee-uh): Single-celled microorganisms, often just called germs. A single one is called a bacterium.

Bernoulli principle (ber-NOO-lee PRIN-suh-puhl): A principle that states that an increase in the speed of moving air or flowing fluid is accompanied by a decrease in the pressure of the air or fluid. Daniel Bernoulli (1700–1782) was a Swiss mathematician who first identified this effect.

biology (by-OL-uh-jee): The scientific study of living things.

blind spot (BLYND spot): A small area on the part of your eye where the retina attaches to the optic nerve that is not sensitive to light.

botany (BAH-tuh-nee): A branch of biology that involves the scientific study of plants.

buoyancy (BOY-uhn-see): The upward, vertical force on an object that is exerted by the fluid in which it is submerged or floating. The **buoyant force** of the fluid is equal to the weight of the fluid displaced by the object.

capillary action (CAP-uh-lair-ee AK-shun): The upward or downward pull on a liquid in a narrow tube, such as inside a plant's stem. Capillary action is caused by **adhesion** (the force of attraction between the water and the plant's stem).

cardiopulmonary resuscitation (car-dee-oh-puhl-muh-nayr-ee ree-suh-si-TAY-shun): Known as CPR, it's a simultaneous forced ventilation of the lungs and repeated pumping pressure on the chest, performed by someone who is trying to sustain oxygen and blood flow in a patient.

celestial body (suh-LES-chuhl BAH-dee): Any object in space that can be studied by an astronomer, including the sun, a moon, planet, comet, or star. Otherwise known as a heavenly body.

chemical change (KEM-uh-kuhl chaynj) or **chemical reaction** (KEM-uh-kuhl ree-AK-shun): A process that changes one substance into another.

chemistry (KEM-uh-stree): The scientific study of matter, especially at the level of atoms and molecules.

Chicxulub crater (CHIC-soo-luhb cray-ter): Located in Mexico near the town of Chicxulub, it's the site of impact of a huge asteroid (or impactor) that struck the earth about 65 million years ago.

chromatography (kroh-muh-TOG-ruh-fee): A method of separating a liquid or gas by allowing it to pass through a material that absorbs each component at a different rate, often in distinct bands and in different colors.

clinically (KLIN-i-klee): Objectively or analytically determined through direct examination.

cohesion (koh-HEE-zhuhn): The force of attraction that holds molecules of a substance together. It's strongest in solids, less strong in liquids, and least strong in gases. Cohesion is what's responsible for surface tension.

comet (KO-meht): A mass of material that travels around the sun in an orbit. When it gets close enough to the sun, some of the comet's frozen material vaporizes, so from Earth it appears as a glowing head attached to a long, gaseous tail. Although sometimes called a shooting star, a comet is not an actual star.

compass (KUHM-puhs): An instrument that tells you which direction is north. It's usually made of a magnetic needle that spins freely on an axis and points north.

compressive strength (kuhm-PRES-iv strength): A measure of how much pressure can be put on an object from the top and bottom in order to determine, with the use of a mathematical formula, how much load the material can bear.

contact force (KON-takt fors): A broad way to describe any force that requires contact between two objects. **Friction** and **tension** are both contact forces.

criminology (crim-uh-NAHL-uh-jee): The study of crime and criminals.

cryptochrome (KRIP-tuh-krohm): A special type of sensory cell in plants, bacteria, fungi—and, many scientists believe, also migratory animals. These cells are sensitive to light, and may be involved in **magnetoreception**.

cuticle (KYOO-ti-kuhl): The outer layer of an **organism** that comes in contact with the environment.

cyanide (SY-uh-nyd): Any of several very toxic compounds that contain a type of deadly chemical that chemists call CN or the "cyano group." Cyanides are produced naturally by plants such as in the leaves of the yew tree, the roots of cassava, and the pits of apples, cherries, and peaches. Cyanides have been used in the mining process and for medical reasons, but also as a method of intentional poisoning.

demospongiae (deh-moh-SPUN-jee-ay): A class of about 8,500 animal species that includes many kinds of sea sponges.

density (DEN-si-tee): A measure of how much matter there is in a certain amount of space.

depth perception (DEPTH per-sep-shun): The ability to judge spatial relationships.

drag resistant force (drag ree-sis-tuhnt fors): A force that slows an object's movement through a fluid (fluid resistance) or through air (air resistance) due to the collision of electrons and ions. Drag force opposes (works against) the forward progress of a moving object. The higher the speed of the object, the greater the drag force. If the shape of the object is properly streamlined, the air will flow more smoothly around it and will cause less drag.

duration (duhr-AY-shuhn): A period of time during which something happens.

electrochemical (i-lek-troh-KEM-uh-kuhl): A way to describe a process in which electricity is produced through a chemical reaction. An **electrochemical cell** is a device that converts chemical to electric energy (or electric to chemical energy).

endoparasite (en-doh-PAIR-uh-syt): A parasite that lives inside its host.

endurance (en–DUHR–uhnts): The ability to continue to do something difficult over an extended period of time.

enzyme (EN–zym): A substance produced by a living organism that causes a chemical reaction.

estivation (es–tuh–VAY–shun): The slowing of activity and often of **metabolism**, usually in hot summer months, similar to **hibernation**.

exponential growth (eks–poh–NEN–shul grohth): An expression that essentially means a quantity that increases surprisingly rapidly. In a technical sense, there's a specific mathematical formula that describes how rate of growth becomes more and more rapid in proportion to the growing total number.

extract (ek–STRAKT): When used as a verb, to draw or pull something out of something else.

fermentation (fer–men–TAY–shuhn): A chemical process where organic substances are broken down or transformed into simpler substances. Manufacturers use the process to make products such as alcoholic beverages, vinegar, and cheeses.

force (fors): In physics, it's something that causes an object to move, change shape, or change speed or direction.

forensics (for–ENZ–iks): A branch of chemistry that studies physical scientific evidence as an aid to medical or legal investigations.

forensic scientists (for–ENZ–ik sy–en–tists): Scientists who investigate physical scientific evidence and establish facts for use in medical or legal investigations.

frequency (FREE-kwen-see): A measure of how often an action or movement repeats within a certain amount of time. **Sound waves** are described in terms of frequency—the number of vibrations a sound makes each second—and are measured in **hertz**.

friction (FRIK-shun): A force that is created when two objects rub together. Friction slows down objects. (See **contact force**.)

ganglia (GANG-lee-uh): A group of nerve cells enclosed by connective tissue, usually located away from the brain and spinal cord.

gas (gas): One of three basic types of matter (the others are **solid** and **liquid**). The **molecules** that make up gas are constantly moving around, and can spread out to fill a space of any size.

genetics (juh-NET-iks): The branch of science that studies the transmission of characteristics from a parent to its offspring.

geology (jee-OL-uh-jee): The branch of science that studies the origin, history, and structure of the earth.

germinate (JUHR-muh-nayt): To sprout or develop.

gravity (GRAV-i-tee): An invisible force that pulls all objects toward one another. The more mass the objects have, and the closer together they are, the stronger the force of gravity. On our planet, gravity pulls objects down toward the center of the earth.

hertz (HURTS): A unit for measuring the **frequency** of vibrations and waves. One hertz equals one cycle per second. The hertz is named after the German physicist Heinrich Hertz.

hibernation (noun) (hy-ber-NAY-shun): A state of inactivity in certain animals during cold winter months that is similar to sleep. To **hibernate** is to enter this state of inactivity. (See **estivation**.)

holothuroidea (ho-loh-thuh-ROY-dee-uh): A class of marine animals commonly called sea cucumbers.

host: (hohst): A larger living thing that harbors a smaller living thing (such as a parasite).

hyperaccumulator (hy-per-uh-KYOO-myoo-lay-tuhr): A plant that can grow in soils that contain high concentrations of metals, and that can absorb these metals through its roots.

impact (IM-pakt): A collision between two bodies that changes their momentum. Also known as an impulse, or an impulsive force.

impactor (im-PAK-tuhr): An object such as a meteor or other celestial body that collides with another object.

imprinting (IM-prin-ting): A rapid response or pattern of behavior by certain species of living things at an early phase of their development, generally to a parent, other living thing, or moving object.

invertebrate (in-VER-tuh-brayt): An animal without a backbone.

kinetic energy (ki-NET-ik EN-er-jee): The energy of an object in motion. The amount of an object's kinetic energy depends on its mass and velocity.

linear momentum (LIH-nee-uhr moh-MEN-tum): The motion of a moving object, measured by multiplying its mass times its velocity.

liquid (LIK-wid): One of three basic types of matter (the others are **solid** and **gas**). The **molecules** that make up liquids can move short distances and take on the shape of the container they're in.

magnetic pole (mag-ne-tik pohl): Either of two opposite areas of the earth's surface where the earth's magnetic fields are strongest. They're located near the North and South Poles. The magnetic poles change their locations slightly over long periods of time. **Magnetic north** is the direction that the north-seeking end of a magnetic needle points.

magnetism (MAG-ne-tiz-uhm): The force produced by a magnetic field, or the force that causes certain materials to be attracted to, or repelled by, a magnet.

magnetoreception (mag-net-oh-ree-SEP-shun): A sense that allows a living thing to detect the earth's magnetic field in order to figure out its location or direction.

mammalian diving reflex (ma-MAY-lee-uhn dy-ving ree-fleks): Something that happens to mammals when they are submerged in cool water. The body's heart rate decreases in order to maximize stores of oxygen and blood flow to the brain.

mandibles (MAN-duh-buhls): The mouthparts of an insect or, in other animals, the lower jaw.

mass (mas): A measure of the amount of **matter** in an object. (See **weight**.)

matter (ma-tuhr): Anything that has **mass** and that takes up space.

medium (MEE-dee-uhm): In chemistry, it's a substance that allows for the transfer of energy from one location to another.

metabolism (meh-TAB-uh-li-zum): The chemical processes inside a living thing that maintain its life.

meteor (MEE-tee-uhr): A rocky object that enters the earth's atmosphere. **Friction** with the air can cause the meteor to heat up and glow in the night sky. Most meteors burn up before they reach the earth's surface.

meteorite (MEE-tee-uh-ryt): A **meteor** that strikes the earth's surface.

meteorologist (mee-tee-uh-RAHL-uh-jist): A person who specializes in the study of the atmosphere and weather conditions.

mixture (MIKS-chur): A combination of two or more substances.

molecule (MOL-i-kyool): A group of two or more atoms that share electrons in a chemical bond and that makes up the smallest unit of a substance.

motor neurons (MOH-tuhr NUHR-onz): Nerve cells along a pathway between the brain or the spinal chord and muscles or glands.

myoglobin (my-uh-GLOH-buhn): A red protein that carries and stores oxygen in muscle fibers.

natural frequency (na-chuh-ruhl FREE-kwen-see): The number of times per second that something **oscillates** (vibrates) when not subjected to an external force.

navigate (NAV-uh-gayt): To find a way toward one's destination.

non-Newtonian liquid (non-noo-toh-nee-uhn LIK-wid): A special kind of fluid that behaves in an unpredictable way; its flow pattern changes depending on the force applied to it, and may become at times more like a **solid** and at other times more like a **liquid**.

nonverbal behavior (nahn-vuhr-buhl bee-hayv-yor): Behaviors such as gestures, postures, and facial expressions, which might indicate how a person is feeling without their use of speech.

nutrient (NOO-tree-ent): A substance that provides nourishment to a living thing.

opposing force (uh-pohz-ing fors): The pushing or pulling of an object by two forces working against each other.

optical illusion (op-ti-kuhl i-loo-zhuhn): An image that uses color, light, or patterns to trick our brains and make us see something that's different from what's actually there.

organism (OR-guh-niz-uhm): An animal, plant, or single-celled life form.

oscillation (o-suh-LAY-shuhn): In terms of sound, it's a repeated pattern of vibration—a steady back-and-forth or up-and-down movement—that happens when something is pushed or struck.

parallel load bearing (per-uh-lel lohd ber-ing): In physics, it describes two or more objects that help support the weight of something.

paralytic (per-uh-LIH-tik): Causing or affected by an inability to move or feel sensations in a part of the body.

parasite (PER-uh-syt): A living thing that lives on or in another living thing (known as a host). Parasites depend on their hosts for food.

penicillin (pen-uh-SIL-uhn): A group of antibiotics produced from a mold called *Penicillium chrysogenum*.

percussion (pur-KUH-shuhn): Striking the surface of something to create sounds or vibrations. Percussion instruments are instruments that make a sound when struck, shaken, or plucked, usually to establish a rhythm.

phospholipid (FOS-foh-li-puhd): A fatty substance that contains the element called phosphorus and is part of a cell's membrane (lining).

physical change (fih-zik-uhl chaynj): In chemistry, it's when a substance is changed from one state—**solid**, **liquid**, or **gas**—to another, without altering the chemical composition of the substance.

physics (FIZ-iks): The scientific study of matter and energy.

phytoremediation (fy-toh-ruh-mee-dee-AY-shuhn): A way of treating polluted or contaminated soil by using green plants to remove or reduce the toxic substances.

phytomining (fy-toh-my-ning): The use of specific plants to help recover metals from soil.

pitch (PICH): The sensation of how high or how deep a sound is. The pitch of a sound wave depends on its **frequency**. A higher frequency (shorter wavelength) means a higher pitch.

polygraph test (PAH-lee-graf test): Also known as a lie detector test, it's a machine designed to detect and record signals in a person's body, such as their blood pressure and breathing rates while they're answering a series of questions, to determine if they're lying or telling the truth.

precipitation (pree-sip-uh-TAY-shuhn): Rain, snow, hail, or sleet that falls to the ground.

prehensile (pre-HEN-suhl): Having the ability to seize or grasp an object by wrapping around it.

pressure (PRESH-ur): The amount of force exerted on a particular area or surface. The amount of pressure depends on the strength of the force and the size of the area upon which that force is applied.

proboscis (pruh-BO-sis): A long tube-shaped part of an animal that is used for feeding.

projectile (pruh-JEK-tyl): It describes something that is shot or thrown through the air. **Projectile vomit** is throw-up that gets propelled violently in one direction.

properties (PRO-per-teez): In chemistry, qualities that help to define or describe something. **Observable properties** are qualities that can be seen, smelled, heard, tasted, or felt. **Measurable properties** are qualities that can be measured, such as an object's height, mass, volume, and density.

psychology (sy-KOL-uh-jee): The scientific study of mental processes and behavior.

refraction (ri-FRAK-shuhn)/**refracted** (ri-FRAK-tuhd): Refraction describes the bending or turning of a wave when it passes through one substance into another, that carries the waves at a different speed. Refracted describes what happens to a wave that is bent or turned.

regeneration (ri-jen-uh-RAY-shun): A way to describe an organism's ability to regrow body parts that have been lost or damaged.

reproduce (ree-pruh-DOOS): Create offspring.

reptile (REP-tyl): A class of **vertebrate** animals that includes snakes, lizards, crocodiles, turtles, and tortoises.

resonance (REZ-uh-nuhntz): A state during which an object's sound (rate of **oscillation**) synchronizes with, and becomes greatly enhanced by, the vibrations of a nearby object.

rhabditophora (rab-dih-TAHF-uh-rah): A class of animals that includes most species commonly known as flatworms.

rhythm (RIH-thuhm): A regular, repeated pattern of movements or sounds.

scyphozoa (sy-fuh-ZOH-uh): A class of marine animals that sting, and that includes jellyfish.

sensory neurons (SENS-ree NUHR-onz): Nerve cells that conduct impulses from parts of the body to the central nervous system.

slipstream (SLIP-streem): An area of air behind a rapidly moving object in which the air moves at about the object's rate, rather than the rate of the still air it passes through.

solid (SOL-id): One of three basic types of **matter** (the others are **liquid** and **gas**). The **molecules** that make up solids can vibrate but cannot move or change places with other molecules.

sound wave (sownd wayv): A kind of energy that moves in waves through air, water, and objects, causing a disturbance from the source of the sound to your ears. The denser the medium, the slower sound will travel through it. Sound waves strike our eardrums and cause them to vibrate. Our brain turns the vibrations into sounds we can hear. Sound waves lose energy as they move (that's why faraway sounds are fainter).

species (SPEE-sheez): a group of living things that are capable of producing offspring with one another.

speed (speed): The distance traveled by an object within a certain amount of time. To be speedy is to move at a rapid pace or rate.

spore (SPOHR): A single cell that can grow into a new living thing without being fertilized by another cell.

steam (STEEM): Water in its gaseous state.

stem cell (stem sel): An unspecialized cell in a multicellular living thing, capable of dividing and becoming a cell with a specific function.

stereopsis (stair-ee-OP-suhs): Depth perception.

structural failure (struhk-chuh-ruhl fayl-yuhr): The loss of an object's ability to carry a load, usually because the structure is stressed beyond its strength limits, causing total collapse.

surface area (SUHR-fuhs er-ee-uh): A measure of the total exposed area inside a given boundary—usually the total areas of the surfaces of a three-dimensional object.

tension (TEN-shuhn): A force that acts on an object to stretch it. (See **contact force** and **ultimate tensile strength**.)

tetanus (TET-uh-nuhs): A disease caused by bacteria that usually enter the body by way of a cut or other wound.

thermal energy (THER-muhl E-nuhr-gee): A type of energy that is contained in the molecules of a substance.

torpor (TOHR-pohr): A state of inactivity, often caused by extreme weather conditions, during which an animal experiences reduced **metabolism**, heart rate, breathing, and body temperature.

 totipotent stem cell (toh-ti-POH-tehnt stem sel): A type of cell that has the potential to develop into any number of different types of specialized cells necessary for a living thing to develop.

toxin (TOX-in): Something that is poisonous.

transmutation (tranz-myoo-TAY-shuhn): The conversion of one chemical element into another. This conversion was long attempted by alchemists without success. Today we know that such a change requires a nuclear reaction or radioactive decay.

true north (troo north): North, as calculated by an imaginary line through the earth at the North Pole.

ultimate tensile strength (ul-tuh-met ten-suhl strength): A measure of the stress needed to break a solid material by stretching it.

vaccine (vak-SEEN): A substance that can be injected into your body that can stimulate it to create antibodies in order to prevent it from developing the full-blown disease.

vampire (VAM-pyr)/**vampirism** (VAMP-pyr-iz-uhm): In folklore, a vampire is a corpse (dead person) that is supposed to rise from its grave at night to drink the blood of living people. In nature, vampirism describes the process in which an animal ingests the blood of other living things.

vaporization (vay-puhr-uh-ZAY-shuhn): A **physical change** from a **liquid** or **solid** to a **gas**, such as through boiling or evaporization.

velocity (ve-LOS-uh-tee): The speed and direction of an object.

vertebrate (VER-tuh-bruht): An animal with a backbone.

vetala (ve-ta-luh): A clever, zombie/vampire-like character from Hindu mythology that has the power to enter dead bodies and reanimate them. Sometimes spelled *vetal* or *baital*.

virus (VI-ruhs): A microscopic, infective agent that can invade cells and cause diseases.

volume (VOL-yoom): 1. The amount of space that an object takes up. Volume is measured in cubic units (length times height times depth of an object). 2. The loudness of a sound.

weight (WAYT): A measure of the force of **gravity** acting on an object. On Earth, your weight is the force of the earth's gravity pulling on you. Your weight varies, depending on where you are in the universe. Weight is a measure of an object's **mass** and the strength of the pull of gravity on the object.

xylem (ZI-luhm): Plant tissue through which water and minerals are conducted upward to the different plant parts.

yield strength (YEELD strength): The highest amount of stress a material can withstand, calculated by mathematical formulas.

zombie (ZOM-bee): A dead person that is revived by magic or sorcery in fictional stories and folklore. In some stories, zombies feed on the brains of the living.

Acknowledgments

I owe thanks to a great many people for their help with this book. No one can be an expert at *every* scientific discipline, and I'm an expert in none of them, so let's just say . . . I'm grateful for all the magical elves who came to my aid.

My deepest gratitude goes to Rusty Davis, a brilliant scientist and veteran teacher. He spent hours patiently explaining complex physics concepts to me in such a way that I truly grasped them.

Thanks as well to Ava Mennin for her engineering help, to Professors Jonathan Losos and Heather Bouwman for their invaluable input and suggestions, to Professor Maria Tatar for her advice with fairy tale origins, to my writer friends Nandini Bajpai, Loree Griffin Burns, Melissa Stewart, April Jones Prince, Kathryn Hulick, and Kate Narita for their multiple readings of multiple drafts, and to my teacher friends Laura Monti, James Lehner, and Amanda Benedict for reviewing science passages. Thanks also to my elementary-school teacher friends Jason Lewis, Kristen Piccone, Kristin Crouch, Linda Bornschein, Yvette Burke, and Amber Kuehler for their help and input and for testing several experiments with their classroom students. Any errors are mine and mine only.

Thanks to my ever-amazing agent, Caryn Wiseman, to Bill Robinson for his dazzling illustrations, to Deirdre Abrams for her professional expertise and input, and to Linda Minton for her keen-eyed copyedits. Thanks to the team at Odd Dot Books, who took a chance on this nothing-if-not-odd book idea: my fabulous editor, Kate Avino; publisher, Daniel Nayeri; editorial director, Nathalie Le Du; creative director, Christina Quintero; art director, Timothy Hall.

Making this book was a true team effort. Many people sprinkled fairy dust on it, and helped transform it into something truly magical.

—Sarah Albee

Odd Dot
120 Broadway
New York, NY 10271
OddDot.com

ISBN: 978-1-250-25761-1

AUTHOR Sarah Albee

ILLUSTRATOR Bill Robinson

EDUCATIONAL CONSULTANT Deirdre Abrams

DESIGNER Christina Quintero

EDITOR Kate Avino

DISCLAIMER

The publisher and authors disclaim responsibility for any loss, injury, or damages that may
result from a reader engaging in the activities described in this book.

Printed in China by 1010 Printing International Limited, North Point, Hong Kong

1 3 5 7 9 10 8 6 4 2